CLIMATE OPTIMISM

CELEBRATING SYSTEMIC CHANGE AROUND THE WORLD

ZAHRA BIABANI
FOREWORD BY CHRISTIANA FIGUERES

CORAL GABLES

Copyright © 2023 by Zahra Biabani.
Published by Mango Publishing, a division of Mango Publishing Group, Inc.

Cover Design: Megan Werner
Cover Photo/illustration: guoquan / stock.adobe.com
Layout & Design: Megan Werner

Mango is an active supporter of authors' rights to free speech and artistic expression in their books. The purpose of copyright is to encourage authors to produce exceptional works that enrich our culture and our open society.

Uploading or distributing photos, scans or any content from this book without prior permission is theft of the author's intellectual property. Please honor the author's work as you would your own. Thank you in advance for respecting our author's rights.

For permission requests, please contact the publisher at:
Mango Publishing Group
2850 S Douglas Road, 2nd Floor
Coral Gables, FL 33134 USA
info@mango.bz

For special orders, quantity sales, course adoptions and corporate sales, please email the publisher at sales@mango.bz. For trade and wholesale sales, please contact Ingram Publisher Services at customer.service@ingramcontent.com or +1.800.509.4887.

Climate Optimism: Celebrating Systemic Change Around the World

Library of Congress Cataloging-in-Publication number: 2022950400
ISBN: (p) 978-1-68481-158-8 (e) 978-1-68481-159-5
BISAC category code NAT011000, NATURE / Environmental Conservation & Protection

CLIMATE OPTIMISM

I dedicate this book to my dad, who has always wanted to write a book. This book carries your love, generosity, and care for me—which has in turn fostered my love, generosity, and care for the planet through the pages. And to my mom, who has always been my cheerleader. Your enthusiasm for life has inspired my enthusiasm for saving the planet.

CONTENTS

FOREWORD___8   MY WHY___11

PART I
WHY OPTIMISM AND WHY NOW?

INTRODUCTION___20

CHAPTER 1
INFORMATION
OVERLOAD___30

CHAPTER 2
NEGATIVITY BIAS___37

CHAPTER 3
THE PRIVILEGE GAP___42

CHAPTER 4
FAILURE OF
IMAGINATION___49

CHAPTER 5
ECHO CHAMBERS___54

TAKEAWAYS FROM
PART I___58

PART II
LOOKING BACK TO LOOK FORWARD

CHAPTER 6
THINGS AREN'T ALWAYS
AS THEY SEEM___64

CHAPTER 7
LET'S TALK ABOUT
EARTH WINS___71

CHAPTER 8
CLIMATE OPTIMISM IN
THE COURTROOM___103

CHAPTER 9
STARTING A
MOVEMENT___129

CHAPTER 10
CHANGE IN
THE MARKETS___148

PART III
CLIMATE OPTIMISM AROUND THE WORLD

NOTE 164

CHAPTER 11
SHE4EARTH–
SOUTH AFRICA 165

CHAPTER 12
PROJECT EIM–
PHILIPPINES 171

CHAPTER 13
ECONOMIES OF PEACE–
PAKISTAN 176

CHAPTER 14
ECO-LUSION–
KENYA 180

CHAPTER 15
PLANTING HOPE–
PALESTINE 184

PART IV
YOUR ROLE

CHAPTER 16
WHAT YOU CAN DO 190

ACKNOWLEDGMENTS 194

ABOUT THE AUTHOR 195

REFERENCES 196

FOREWORD

BY CHRISTIANA FIGUERES

Optimism is not the result of a success; it is the necessary ingredient to address a challenge if we want any chance of success. Optimism is not blind to reality. It is not oblivious to the hard facts. Optimism is a choice, and I have applied it many times as the input to tackling my personal and professional life challenges. The historic Paris Agreement on Climate Change would not have been achieved if my colleagues and I had not intentionally applied a stubbornly optimistic approach to the negotiations, even when most people thought it was impossible to get all the countries of the world to agree, and we had no way of knowing whether we would succeed.

This book is a brilliant, down-to-earth reminder about the importance of that mindset today. It shows us that distraction, doubt, and despair are heavy burdens we can no longer afford to drag, that transformational change in the world starts with each of us and how we show up in the world.

Underneath the grief and the anger we all harbor a hunger for healing. Zahra's stories, and her keen grasp of the lesser-known trends within environmentalism, show us that healing is possible: it starts with bravely facing the hard facts of the climate crisis and humbly moving toward reconciliation and restorative justice.

When I look to the future, I think of how my goddaughter Zoe, who is eleven years old, and her peers will look back at this moment and be appalled at how close we came to losing the Amazon, the ice caps and glaciers, and the coral reefs. I expect she and her friends will laugh at how ludicrous our old-fashioned ideas were: digging up ancient fossils just to burn them, clearing natural forests to just grow other plants, or killing animals just to eat them. They will be incredulous that our cities were once filled with loud vehicles and toxic fumes, that we scraped the sea bottoms and paved over wetlands.

But my goddaughter and her peers will also marvel at all we collectively achieved in such a short space of time when we finally woke up to the consequences of our doings. They will be amazed at how we transformed entire systems for the benefit of everyone, especially the most vulnerable.

And they will be grateful for all the disruptors, of all stripes, sizes, and ages, who pushed and pulled us, inspired us and admonished us, and ultimately helped us to move into responsibility at the speed and scale we needed.

The path to that future may seem difficult or perhaps even impossible from where we stand right now. But impossible is not a fact, it's an attitude, only an attitude, and our attitudes are firmly under our control.

The fact is, we can choose to change our policies, our technology, and our financial levers to protect life on earth. And we can do so powered by the conviction in our capacity to effect those changes and our capacity to heal. We can also be continuously inspired by the extraordinary efforts of so many of our fellow citizens around the world who are already working so hard for change, even as obstacles keep blocking the way.

Optimism is a candle in the darkness. It can brighten the minds of each one of us if we choose to light it. And as the Buddha said: brightness of mind is both the final goal of the path of enlightenment and also the first step. A bright mind is how we can proceed and meaningfully contribute to restoring the incredible web of life we are so fortunate to be part of.

Christiana Figueres
Founding partner of Global Optimism
Former executive secretary of the United Nations
Framework Convention on Climate Change (2010-2016)

MY WHY

> **Climate optimism:** A framework based on the idea that we can restore the earth back to health, and in doing so protect the people that inhabit this planet.

A small percentage of humanity has slowly been poisoning our shared home, from the polluted groundwater below our feet to the tainted skies above our heads. For some, it's easier to give in to the poison, accepting it as the new normal. For others, this poison has infiltrated deep into their countries, homes, families, and bodies. In a few years, it will be impossible for anyone to avoid the damage that this poison carries with it: increasingly intense natural disasters, more pandemics, hotter temperatures, and wars over depleting natural resources. So why pick up a book on climate optimism?

This book is for young people who are terrified for their futures.

This book is for climate advocates who tirelessly call for action but are run down from fighting what often feels like a losing battle.

This book is for elders who have been fighting for change for as long as they can remember.

This book is for my future children, who are not responsible for the world they will be born into, yet will reap the consequences of our actions if we fail to act rapidly.

This book is for a seventeen-year-old me, impassioned to "save the planet," yet overwhelmed on how to do so.

At the risk of coming off as exclusionary, let me also state who this book is *not* for.

This book is not for the political leaders around the world who have failed to heed scientists' warnings and take adequate action to protect the planet and its people.

This book is not for corporate leaders who value sustainability as an afterthought and whose prioritization of profit over people and the planet have gotten us into this ecological disaster.

This book is not for members of civil society who are lukewarm on the issue of the climate crisis, claiming that it is a problem, but far less of a problem than scientists agree on.

Although it is in my best interest to sell as many books as possible, I want to make sure that nothing in this book gets misinterpreted, twisted, or abused by those in power who use my optimism as an excuse for inaction or delay.

I wrote this book because I believe that, in light of our dwindling resources, hope is an unlimited resource.

I believe that climate optimism is necessary for mobilizing change. I believe that climate doomism—a fatalistic view of our future centered on the belief that the earth is too far gone—is now so pervasive that it rivals climate change denialism. As said by Andreas Karelas, author of *Climate Courage*, "people will choose not to believe in climate change if they believe there is no way to solve it."

I believe in the inextricable link between climate change and any social justice issue. To solve the climate crisis, we must actively work to dismantle systems of oppression that uphold injustices.

I believe in empowering and funding communities of color and Indigenous communities who are facing the brunt of consequences of climate change despite contributing the least to the issue.

I believe in love, the greatest of all things. When our love for our planet and its people supersedes our love for profit, earth will begin to heal.

I believe that good people can believe bad things and that climate change deniers are not inherently bad people; they just have been misinformed by the media and networks of disinformation.

I believe that we need to convince everyone of the urgency of the climate crisis to drive the political will and reform necessary to make a change.

I believe that *you* have a role in fighting the climate crisis and that if everyone who cared about the planet got involved in high-impact movements like those featured in this book, we would save our planet from catastrophic warning.

Finally, I believe our propensity to pioneer solutions is far greater than we realize. Movements within the larger environmental movement have the power to bring about great change for both the planet and its people. Though there is ample room and demand for involvement in these movements, there is a lack of awareness of them, as our dialogue around environmentalism too often focuses on sensationalized headlines and political quagmires.

ICYMI (in case you missed it) on the cover page, I will remind you of my name, who I am, and why I am writing to you.

My name is Zahra Biabani, and I am a youth climate activist from the United States (Houston, Texas, specifically).

I tell most people that I started my climate activist journey at a Climate Reality Leadership Corps Training in my hometown in 2016. This was my first formal foray into the climate space, but truthfully, my journey began in a less-than-wholesome way ten years prior.

At the start of each new month, the newest edition of *National Geographic Kids* would faithfully arrive in our mailbox. Something about this issue stirred me into action like no other had before. Moved by the compelling images of polar bears losing their homes to the greed and apathy of humans, I decided to go door-to-door in my neighborhood, telling people about their diminishing homes (melting ice caps) and collecting donations in one of my mom's old shoeboxes (they were Dr. Scholl's, I think).

My eight-year-old brain clearly did not hold onto that passion project for long, as a few years later I discovered the shoebox with fifteen dollars in donations in a corner of my closet at the end of the school year. Yep, little me accidentally stole money under the guise of charity work. I blame my short attention span.

After a short stint of obsession with animals, the latter bit of elementary school through late high school was marked by a commitment to various social issues. I made videos about mass incarceration, started the Democrats club at my high school, and went door-to-door campaigning for progressive political candidates with my mom, as I was too young to do so by myself.

I assumed that my interest in the melting ice caps had waned, but it wasn't until I realized the connection between social inequities and climate change that my passion for preserving the planet fully flowered.

In the summer of 2017, I worked at a Social Services agency. I was volunteering at our food bank on an afternoon where the sooty clouds loomed low and the sky seemed to be grumbling at its mere existence. I remember a fellow volunteer speculating that despite the projections from meteorologists, this storm would be the same as all Houston summer storms—temperamental, sudden, and brash—but far from devastating.

He was wrong.

Hurricane Harvey touched down in my hometown on August 25, 2017, bringing an inexplicable amount of destruction. Up to nine months later, families (mostly Black and Brown) who came into the Social Services office were still bereft of housing, FEMA support, and the remnants of their lives pre-Harvey.

It was in the agency's aged office chair that I firsthand heard testimonies of people suffering from this environmental catastrophe; a young woman who had a double miscarriage due to household mold from the flooding during the storm, an old woman who now had panic attacks every time it rained, and families who were already destitute getting knocked down even further.

Knowing the connection between natural disasters and a changing climate, I saw firsthand how climate change exacerbated social inequalities. So the next year when I started university, I decided to study environmental sociology and environmental science. It was at this time that I began to share what I was learning both inside and outside the classroom on social media.

And here I am, two years post-grad, sharing what I am learning to democratize climate education and spur climate action. I have cultivated a community of 50,000 advocates who receive weekly positive news stories about the planet, given a TED Talk on climate optimism, and started a company to advance the accessibility of sustainable and ethical fashion. Writing this book has been my greatest accomplishment of all. Between the research and conversations that have culminated in these words, I have learned so much and gained a newfound sense of hope for our planet, and I can only pray the same is true for you.

—

I am writing to you because I believe that our nihilism toward environmental progress is one of the greatest threats to the planet herself. In my optimism research, I've read several incredible books that detail progress and prosperity in the last few decades. These books display aesthetic graphics of decreasing infant mortality rates and extreme poverty, amongst other encouraging trends that we will talk about later.

It's rare to find such a graph detailing environmental progress, but it's no wonder why—environmental progress has been slow, difficult to pinpoint, and laborious to find. In recent years this progress has been mangled in political polarization and populism, fueling inaction.

As difficult as it may be during these times, I believe that by looking for the good, we enable ourselves to be agents of good in the most important fight of our time. Reflecting on my experiences and what I have learned from others' activism journeys, I know that hope is a self-fulfilling phenomenon. We find hope by observing what is going on around us and cultivate that hope by getting involved in what we observe.

My goal for this book is to awaken you to the seeds of change that are hungry for resources and push you toward an enduring hope that can be gained through the practice of nurturing and eventually propagating these plants.

I feel lucky to speak to you through my writing. Thank you for your willingness to hear my thoughts; I pray they inspire you.

May these words lift you out of your pits of hopelessness and despair, and empower you into action. May this book show you that climate solutions are being pioneered all over the world, by the people who have contributed the least to the climate crisis. May these stories inspire you, leading you to pursue your own climate work.

PART I

WHY OPTIMISM AND WHY NOW?

INTRODUCTION

—

"It's a tough time to be alive," said someone, probably, on Twitter fifteen minutes ago; and someone in March of 2020, just as COVID-19 was breaking out worldwide; not to mention someone living through the Black Plague, recording their thoughts on stretched goat skin.

Whenever I find myself lamenting the abysmal state of our world, I try to picture what my life would look like if I was born eighty-eight years earlier, in 1930. I would be born into the Great Depression, perhaps in a shantytown alongside other destitute families. Sustainability is an integral part of the culture, but by way of necessity and frugality rather than stewardship. As a Brown person, life is even more difficult, with the continuous threat of racial violence and deportation.

Just as things start to pick up, they head downward. At the age of ten, war is on the horizon. My undeveloped brain cannot wrap my head around why America would be threatened by countries so far away, but I acknowledge that perhaps President Roosevelt knows something I don't. So I bid my father goodbye, as he is recruited for the draft. In 1941, the Second World War breaks out.

Talk about tough luck.

And what if I was born in the Stone Age—forced to migrate from temporary home to temporary home in a frigid ice-age environment? Sounds horrid. Of course, I'm comparing this to the only reality I know. As I write this, I'm sitting in my ergonomic chair with my double monitor and Spotify "Music for Writing" playlist humming in the background, so it is no wonder I'd prefer my comfortable reality to that of someone living in 10,000 BC.

This comparison seems like an easy exercise. And yet, when we look at the rhetoric surrounding us, few people find hope in how far humanity has come.

Instead, we tend to reflect on the "good old times," an invocation to a bygone, imprecise time when things were presumably easier. Positive memory bias, also known as rose-tinted glasses, engulfs our personal and collective memories, making the past seem inherently better than the present—especially in times of stress.[1]

This bias makes it easier to amplify any negatives we face in the present, while minimizing the positives. It's easier to claim that the world is getting worse and that humans are getting worse with it. And it's even easier to do so on social media, where this short-sighted negativity spreads like wildfire.

Perhaps you infrequently put on a pair of rose-tinted glasses, preferring to stick with polarized UV-protected lenses instead. These safeguard you from the false belief that things in the past were necessarily better than those of the present and allow you to fix your focus on the here and now. But you still might be vulnerable to the idea that the here and now is fundamentally different and unique than any other time in history—the unprecedented bias.

In the last few years, the term "unprecedented" has become a bread-and-butter staple in conversations—rising through the dictionary ranks for the title of People's Choice 2020 Word of the Year by Dictionary.com.[2] And in 2020, our collective vernacular deployed the word in attempts to explain everything from the pandemic to police brutality to attacks on the Capitol building of the largest democracy in the world.

Despite the word's connotation implying a once-in-a-blue-moon use case, it is a term that has never gone out of style. Usage of the word was first documented in 1795 and has been increasing ever since, according to an analysis done with a Google-powered textual analysis tool.[3] Things have always seemed insurmountable, once-in-a-lifetime, and extraordinary.

Simply understanding how humanity has collectively experienced and interpreted events of our past helps us to better understand the ostensibly unassailable challenges of our present. And with knowledge of where we have been, we can better understand how we have improved and how we can use learnings from our past to shape our future.

Although many would argue that it is not useful to understand suffering and current events through relativism, I think doing so is underutilized and much needed in today's day and age—when nihilism stemming from fear of the state of our world is rampant and rhetoric around the end of times is equally pervasive.

To overcome the human tendency of embellishing the past and viewing the present as exceptional, it is helpful to look at history through a statistician's lens.

In my lifetime, and likely yours, much of the world has experienced a relatively constant era of growth under the liberal world order and globalization. Despite the flaws of globalization, a non-exhaustive list of which includes the deterioration of environmental and labor rights, widened inequality gaps, and the creation of the outrageous billionaire class, it has indisputably contributed to a decrease in poverty, increase in access to goods and services, and lifesaving and changing innovations.

- ◊ **Extreme Poverty:** The percentage of people worldwide who live in extreme poverty shrank from around 37 percent in 1990 to 10 percent in 2015.[4]

- ◊ **Life Expectancy:** Life expectancy has increased by more than six years between 2000 and 2019—from 66.8 years in 2000 to 73.4 years in 2019.[5]

- ◊ **Literacy Rates:** In the last sixty-five years, the global literacy rate has increased from 42 percent in 1960 to 86 percent in 2015.[6]

- ◊ **Infant Mortality Rate:** Globally, the infant mortality rate has decreased from an estimated rate of sixty-five deaths per thousand live births in 1990 to twenty-nine deaths per thousand live births in 2018.[7]

◊ **Better Access to Water:** Across Low- and Middle-Income Countries (LMICs), access to improved water overall increased between 2000 and 2017. For piped water, the safest water facility type, access increased from 40.0 percent to 50.3 percent. Access to sewer or septic sanitation and improved sanitation overall also increased across all LMICs during the study period. For sewer or septic sanitation, access was 46.3 percent in 2017, compared with 28.7 percent in 2000.[8]

◊ **Access to Electricity:** In 1990, around 71 percent of the world's population had access to electricity; this has increased to 87 percent in 2016.[9]

◊ **Death Rates from Air Pollution:** Since 1990, global death rates from air pollution have nearly halved.[10]

◊ **Democratic Governance:** At the end of 2017, 96 out of 167 countries with populations of at least 500,000 (57 percent) were democracies of some kind, and only 21 (13 percent) were autocracies. The remaining 28 percent were pseudo-democracies with elements of both democracies and authoritarian states. The share of democracies has been on an upward trend and is just shy of its post–World War II record (58 percent).[11]

The Legatum Prosperity Index, which tracks prosperity across the globe through open economies, empowered people, and inclusive societies is a testament to this. 2019 marked the highest level of global prosperity ever. Even during a global pandemic, in the following years of 2020 and 2021, Global Prosperity levels plateaued.

The connection between these metrics of prosperity and globalization is apparent. Globalization necessitates the exchange of ideas, goods, and services that contribute to open economies, more empowered people, and societies with a greater acceptance and tolerance.

To be clear, the same prosperity opportunities have not been accrued to everyone equally. The people and descendants of people Western civilizations like the United States were built on—Indigenous peoples, Enslaved peoples, and other minority communities—have been continually oppressed under systems disguised as justice (e.g., the criminal justice system). Likewise, communities in the Global South have been continually exploited under systems disguised as aid (e.g., much of the international aid system). The lasting impacts of these systemic injustices are apparent in the unabating levels of income inequality.

Even amid the rife inequalities that plague our world, when we look at the statistics and growth trends (which we will get into soon), it is clear that by many accounts, things are better than they ever have been. Perhaps it might be more accurate if we rephrase it to, "It's a tough time to be optimistic."

But if we look at the statistics on the climate crisis, it is also clear that this behemoth problem threatens to upend this era of prosperity in a way that nothing has ever done before.

By some estimates, climate change could spur the greatest refugee crisis in history—with an estimate of 1.2

billion people being displaced by climate-related events in the next three decades;[12] incite one of the biggest economic downfalls—costing the world $178 trillion over the next fifty years;[13] and cause a large-scale public health crisis—with medical professionals from across the world coming together to say that climate change might be the biggest threat to public health.[14]

The Department of Defense has called climate change a great "threat multiplier"[15] due to its capacity to weave into and amplify every single crisis, making each one worse while simultaneously getting worse itself.

This phenomenon should make us worried about the state of our planet, but we cannot allow it to debilitate us.

Because of the magnitude of the threat that climate change poses to humanity, my message of climate optimism is occasionally met with disdain and criticism. Some watch my videos and leave comments about the irreversible effects of climate change and how my "pointless videos" do nothing to reverse the damage that is being done. They claim my celebration of the positive does nothing to stop corporations from wreaking havoc on the planet and its people. I get it. It is hard to be optimistic when it feels like the world is falling apart, being ripped open at the seams by the very people who have promised to protect it.

Let me be clear: climate optimism is not the expectation of a salvaged planet. Instead, it is the proclamation of hope for a healthier and more just planet and the pursuit of actions aligned with what needs to be done to get there.

Those comments pale in comparison to those I encounter expressing relief and comfort at the sight of positive climate news. The majority express that they look forward to Fridays all week long, knowing they can find a feel-good video on my page and an email in their inbox where I present the "earth wins" of the last week. Success of this series has undoubtedly been driven by demand for it.

In the past year, the demand for coverage of good news has skyrocketed to levels that David Beard, the executive director for newsletters at National Geographic, says is "unlike anything he's seen before."[16]

Anecdotally, people speak on the need and benefits of positive news. Empirically, this has been validated by data suggesting an inverse relation between positive news digestion and feelings of anxiety.[17]

Even the World Health Organization (WHO) emphasized the importance of positive news in dealing with COVID-19. During the initial outbreak of the pandemic, the WHO Department of Mental Health and Substance Use released a paper that advised people to "find opportunities to amplify positive and hopeful stories."[18]

With media coverage, when demand increases, supply follows. Since I began reporting on positive environmental news, I have seen a significant uptick in the number of brands, nonprofits, and news outlets similarly highlighting good climate news.

I am certainly not clicking my heels, feel-good all the time. Before diving into climate optimism, I found myself sinking into the quicksand of nihilism, overwhelmed by what I heard on the news and what I learned in school, much of which was discouraging. I knew I wanted environmentalism to be my life's work and to ensure it could be, I needed to find a way to sustainably engage with the news around me.

Climate optimism allowed me to reacquaint myself with the work I do and the "why" behind it. This is not to say I do not experience days where I am writhing in eco-anxiety.

> **Eco-anxiety:** Extreme worry about current and future harm to the environment caused by human activity and climate change.

Whenever I feel dejected, crumbling under the weight of projections for rising temperatures and species loss, I remind myself that by several accounts, the world is better than it has ever been.

And though our window to act is narrowing, our portal of opportunities is expanding.

We are still facing several overlapping and intersectional global crises, all amplified by the climate crisis, but humanity is more advanced, connected, and well-off than it has been at any previous time in history. When it feels like the world is ending, these reminders help to ground us and move us forward. This hope is what I aim to provide others with as well.

Some call this mindset toxic positivity, but for me, it's far from toxic; it is empowering, enabling me to make the change I want to see in the world (thanks for the quote, Gandhi). At worst, toxic positivity is a cursory dismissal of important issues blanketed in a lack of empathy. At best, it's an earnest, albeit callous, attempt to soothe over real pain with the notion that "at least something worse isn't happening."

There are several reasons the framework of climate optimism is difficult for people to accept. This section will attempt to unpack why this is such a hurdle, uncovering the ways we have been primed to fixate on the negative; and in doing so, prevent ourselves from pursuing the positive.

CHAPTER 1

INFORMATION OVERLOAD

"INFORMATION IS NOT KNOWLEDGE."

—ALBERT EINSTEIN

In college, I frequented a method of time management called the Pomodoro Technique, invented by a man I have never met but to whom I am forever grateful, Francesco Cirillo.[19] This method, popular among those of us who get distracted,[20] allots twenty-five minutes of uninterrupted, dedicated grind time followed by a five-minute break. Pretty good deal if you ask me.

I recall being in a quiet study room working on a formal logic problem set. After my twenty-five grueling minutes of focused work, I opened TikTok on my phone and a *New York Times* article on my laptop (the memory stuck because my headphones were not in and alas, the TikTok audio shocked the members of the silent floor, who proceeded to shoot me less-than-friendly side-eyes). Surely, I thought, I could be present for both activities, despite knowing that multi-tasking is proven to be ineffective, especially in today's world where information is everywhere and companies are constantly fighting for our attention in both subtle and not-so-subtle ways. To some extent, we are all victims of information overload...but how we wrestle with it is up to us.

Widespread access to the World Wide Web has opened the gates to unlimited information, a powerful tool that can both mobilize and paralyze.

Information overload is something Gen Z and Millennials are facing to an unprecedented degree. In 2020, internet users created an average of about 1.7 MB (megabytes) of data per second.[21] Assuming a screen time of around eight hours a day for Americans in 2020,[22] that comes out to around 48,960 MB per day in data usage, converted to 48.9 GB (gigabytes)—more than a typical iPhone storage of 32 GB (which, for reference, is around 16,000 images taken on an iPhone). If you have not thought of yourself as a content creator before now, maybe that statistic will have you thinking differently.

It is projected that by 2025, people will be generating around 463 exabytes of data per day.[23] That equates to a *lot* of iPhones. Internet access is on the rise outside of just the Western world. As of January 2021, there are 4.66 billion internet users around the world, spending an average of seven hours per day on the internet across all devices.[24]

When bombarded with information, especially information that is framed negatively, it understandably becomes difficult to break free of the temptation to keep scrolling.

Consequently, the anxiety evoked by this onslaught of oftentimes gut-wrenching information can trap us in action paralysis. It is easier to keep scrolling through your "For

You" page than acting on a systemic issue that seems too large to grasp, like climate change.

Information warfare, the act of using information or false information to advance some goal, is empowered by information overload. The mass amount of information disseminated daily makes it easier for individuals, governments, and companies to spread disinformation. It is important to specify the differences between disinformation and misinformation, two buzzwords that have been flying around social discourse in the last few years.

Disinformation is defined as deliberately deceptive false information.[25] The popular girl from high school spreads a rumor about another girl to increase her chances of winning prom queen (my head is clearly stuck inside an early 2000s movie)—that is disinformation. Misinformation is the spreading of false information, absent the intention to mislead.[26] Grandma posts a graphic (noticeably absent of citations) in the family group chat about the danger of GMOs out of care for your health—misinformation.

Our ability to differentiate information from mis/disinformation is hindered by our shrinking attention spans (thank you, TikTok). I have fallen victim to the TikTok attention span shortage, sometimes finding myself so unable to commit attention to a single form of media that I attempt to read articles while scrolling through the app (hint: it doesn't work).

The intensity of these phenomena was affected by the events of 2020. TikTok downloads skyrocketed at the outset of the COVID-19 pandemic. In the first quarter of 2020, when the pandemic began, there were a total of 315 million

TikTok downloads worldwide, a 58 percent increase from the previous quarter.[27] Personally, I refused to download the app for the first few months of quarantine, convinced that it was too saturated with people half my age twerking and lip-syncing. Fast forward a few years and here I am, dancing alongside them (and spreading positive climate news while I'm at it).

Lockdowns and quarantines drastically increased screen time, as you might have guessed. In a cross-sectional study of adolescents, screen time increased from 3.8 hours/day pre-pandemic to 7.7 hours/day during the pandemic.[28] Kids were glued to their devices for everything from school to leisure to community. Life began to exist in the digital space for many young people during this time.

Digital activism flourished in this time of lockdown, allowing people to mobilize and feel a sense of activation for a cause from inside the confines of their childhood bedrooms. Sharing a post or retweeting a tweet became an accepted—if not promoted—form of advocacy.

The resurgence of the Black Lives Matter movement following the brutal murder of George Floyd was, in large part, fueled by the rising screen times and devotion to one's social media.

Activists (including myself) and other educators began to use social media to spread information about important issues, using infographics to aesthetically synthesize scientific studies on workplace discrimination or particulate matter concentration in communities of color. And although this landscape has slightly shifted since the start of the pandemic, the relatively nascent form of communication, mobilization, and advocacy has remained.

The landscape cultivated by this social media generation is speckled with an assortment of shallow holes.

Each hole highlights an issue of great importance—keyword: highlights—be it the phenomenon of hidden racial biases, environmental injustice, or institutionalized discrimination. Scrollers jump from hole to hole, gathering a surface-level understanding of an issue before moving on to the next one, and from time to time coming across a selfie of a former high school acquaintance. Consumers of content begin to think of themselves as relatively informed on an issue or subject after watching a thirty-second explanatory video.

Social media platforms reward users with dopamine rushes so that they stay on the app, inhibiting them from doing their research after consuming any educational content or getting off the couch to do something about an issue they learn about.

Older generations are not immune from information overload, either. Facebook, a frequented site for older generations, has fallen under scrutiny for its role in spreading misinformation and disinformation. The platform's outsized role in spreading misinformation online was perhaps most evident during the 2016 elections, when the top-performing fake election news stories on the platform generated more engagement than the top stories from major news outlets, according to an analysis.[29]

With an average of 1,500 posts eligible to show on a user's News Feed each day[30] and 31 percent of Americans regularly getting their news from Facebook,[31] it is no wonder that misinformation and information overload are active threats on this application.

And with the click of a button or the tap of the screen, scrollers can share these shallow holes of information—whether factually correct or not—on their feeds, thereby burdening the news feed and making it more difficult for people to sift through accurate information.

In this way, the proliferation of information has led to a paucity of attention.

When so much of this information is negative, it becomes difficult for people to believe that any good is happening in the world. On its own, information overload can overwhelm. When compounded with the negativity instinct, which we will explore next, information overload has the debilitating power to trap people in sinkholes of "doomism."

OVERCOMING INFORMATION OVERLOAD

To avoid the risk of echoing your mom's urge to "get off your phone," I want to provide three tangible tips to help you resist the information overload phenomenon.

1. **Know when to stop "doom scrolling."**
    ◊ Set time limits on your media and news apps. Flag stories that you want to read or watch later (through

an application like Pocket) and stick to an amount of screen time/reading time that feels right for you.

2. **Avoid misinformation/disinformation.**
   ◊ Hold your sources accountable for proper and robust citations.
   ◊ Whether it be a fifteen-second video or a *Washington Post* article, content creators are responsible for citing where their information is from.

3. **Critically assess intentions.**
   ◊ When consuming information, it is important to ask yourself the provider's intentions of the information. Do they stand to benefit from making you think a certain way or believe a certain thing? This practice helps you to narrow down your media sources and combat disinformation/misinformation.

CHAPTER 2

NEGATIVITY BIAS

"IF IT BLEEDS, IT LEADS."

—OLD JOURNALISM ADAGE

As I nestled into seat 32B, on a flight I regularly took in college from Nashville, TN, to my home in Houston, TX, I began to think about the work it takes to prepare a plane for takeoff. Almost every time I fly, I am awe-struck by the sheer ability of me and dozens of strangers to fly tens of thousands of feet in the air, momentarily bound to each other by the laws of physics and a shared trust in the pilot. Few people know the physics and mechanics that allow planes to work but most trust them anyway.

There were an estimated 25,700,000 flights worldwide in 2021, only twenty-six of which resulted in accidents.[32] These successful flights did not get airtime, but the ones that failed most certainly captivated audiences all around the world, leaving them glued to their TV screens in a combination of sorrow and disbelief.

Fewer incentives exist for media platforms to focus on the good than the bad. Think about a time when "Breaking News" was about a positive occurrence. I'll wait...

The reasoning for this may seem obvious; it is undeniably important to report on anomalous tragedies to potentially prevent them from happening again. As long as injustice exists in the world, we have a duty to bring it to light. At the same time, the tendency for media outlets to perseverate on negative news can trap viewers in a cycle of cynicism and frustration.

This tendency is exceptionally clear in the media's coverage of climate change. A study of 350 articles on climate change across six countries showed that disaster framing was employed in 82 percent[33] of articles. Perhaps more shocking is that disaster framing, compared to other kinds of framings that were being assessed, was most often used in the headlines and first few lines of articles—often the most salient parts of an article.

A transition away from casting doubt on climate projections in the media is significant, and long overdue. But it is important to balance reporting the real threats alongside the real solutions that society can engage in and support to reduce these threats.

As a result of the lopsided coverage of climate disasters over climate solutions, consumers of media are quick to name climate consequences but cannot do the same with climate actions.

This is explained by the availability heuristic— the tendency to recollect information based on recent experiences.

It's not all on the journalists or media outlets. Despite the rising demand for positive news, humans have some sort of innate proclivity toward shocking and negative news.

Because of a phenomenon that psychologists call the negativity bias, we are hardwired to be extra attentive to stories that evoke negative emotions. In a 2014 McGill University study, a pair of communication and data science experts used baseline eye-tracking to see the articles participants read when presented with a selection of stories (both those that were more positive and those that were negative) from a news website. The results showed that, when surveyed, despite a majority of the people claiming that they preferred positive news stories, a majority actually were more drawn to negative-skewing stories.[34]

Regardless of how badly we yearn to stay positive, we are hardwired to be far more attentive to negativity.

Decades of scientific literature point to the existence of the negativity bias in decision making, learning, perceived importance, and identifying truthfulness. In other words, information with a negative frame weighs heavier in decision making,[35] stimulates faster learning,[36] is more compelling to readers,[37] and receives significantly higher truth ratings than equivalent statements framed in a positive manner.[38]

The negativity bias is believed to serve evolutionary purposes.[39] It staves off dangerous threats by inflating the risk of those threats and making us extra attentive to them. We don't sleep with one eye open to bear witness to the beautiful sunrise that paints our shared sky each morning—we sleep with one eye open to protect ourselves from any potential enemies.

When the overwhelming tide of the information overload is paired with our negativity bias, it becomes nearly impossible to dig ourselves out of despair.

Fascinatingly, the negativity instinct is a phenomenon that fervently spreads in certain subsets of the population. Individuals with the time and mental space to read the morning news every day and flip on the evening news each night are likely privileged in several ways. Electricity, access to impartial news outlets, and uncensored content are privileges many people are not afforded. In this way, privileged individuals and communities that do not have life-altering crises to worry about often are the most frequent victims of the negativity bias.

The next chapter will reveal how our privilege compounds the negativity bias, resulting in a gap between reality and our perceptions.

OVERCOMING NEGATIVITY BIAS

The tired mantra urging people to "just stay positive" will not cut it for overcoming the deeply entrenched negativity bias that is enforced by the media and our brains. Here are some tips on how to conquer it in a manageable way.

1. **Follow positive news sources.**
    ◊ Research clearly shows that negative news spreads further and faster than positive news on social media, due to the preferences (and subconscious biases) of readers and viewers. By purposefully seeking out news sources or platforms that try to highlight good news stories, you are combating the negativity bias.

2. **Look at the "Trends" in Part II of this book.**
    ◊ If you are reading about how bad something is now, it is helpful to look back and see if any progress has been made throughout history. For example, if looking at infant mortality rates, look for statistics regarding the trajectory of these rates throughout time. This exercise does not always point to positive trends (as we know, almost all measures of climate change are getting worse), but it does help us view the data more holistically.

CHAPTER 3

THE PRIVILEGE GAP

—

"THERE IS NO THEM, ONLY US."

—U2

My parents took me on a trip to Zambia when I was in the fifth grade. I was bummed about going to a far-off country that I had never heard of and especially jealous of my friends spending their spring break at Disney World. At the time, going to Epcot's Morocco was more appealing than going to Morocco's Morocco.

As my parents hoped, the trip was an eye-opening experience for me—exposing me to a world so different from mine. I remember journaling about the transformative experience and how I would never complain about not getting toys ever again (spoiler alert—I didn't keep that promise). In the long term, this visit drastically changed my worldview, but the short-term changes I thought I would undergo were in retrospect quite ignorant.

After returning to the states, I spent extra hours on FreeRice.com, diligently trivia-ing for the provision of grains of rice for under-fed children in Africa and taking extra care to minimize my food waste (and by extra care I mean I hid my half-eaten peanut butter and jelly sandwiches under my bed instead of throwing them away, resulting in a bust a few weeks later when my parents realized my bedroom smelled like moldy grapes).

During the trip I met many people who went without food and other vital resources, and I pledged to use my birthday money to fundraise for communities in Africa, the actual location of which I did not know. At that young age, I viewed Africa as a monolith of under-resourced countries and full of suffering people. Problematic, I know. But not different from how many adults still conceptualize countries in the Global South (regions south of the equator outside Europe and North America that, due to colonial and extractive capitalist forces, struggle with systemic poverty).

With great privilege comes great ignorance.

Regardless of how much you read about or even travel to different countries, unless you are consistently living in a place, you cannot truly know what it is like to exist there. The privilege gap oftentimes results in a distorted perception of how "other" people exist and thrive. It exists within communities, cities, countries, and across international borders.

I was quite young when my mom told me she had lived in Tanzania in her youth. I was shocked and mistakenly pictured her and my grandparents huddled in a hut without air conditioning or a stove to cook on. This was not her reality and she did, in fact, have a functioning stove.

We tend to label people and communities we do not know or identify with as "other." We amplify the problems we see elsewhere, classifying them as problems unique to "faraway places" while minimizing the problems we face in our communities. The lack of access to clean water is not only a problem in South Sudan, but also in Flint, Michigan, USA. Air pollution is not only a threat to communities in Mumbai, but also in Compton, California, USA, and Milan, Italy.

This concept of "othering" has been defined as a product of post-colonial theory and is a sociological term commonly used to explain how racism and other forms of oppression are justified.[40]

When we "other" people, it's easier to treat them differently. And since we are humans with the instinct to protect ourselves and our kin, differential treatment oftentimes leads to discriminatory treatment. The usage of the "separate but equal" doctrine is a prime example.

Although my eight-year-old self did not know any better when learning about Africa for the first time, many adults are guilty of this practice and perpetuate the phenomenon to the detriment of people all over the world.

Othering has been embedded into the way many Western countries engage in international relations, with much of foreign policy historically being dependent on American or European exceptionalism, imposing their values onto other countries based on a fundamental belief that what they do not understand, the "other," is worse than their own norms. This dangerous belief system was evident in an over-confident and over-zealous American invasion of Iraq, which was informed by little information to back up the monumental decision, a cursory analysis of the evidence, and an unplanned strategy for helping the Iraqi people. In this way, exceptionalism is often imperialism masquerading.

The late Hans Rosling published a book called *Factfulness* that looked at common cognitive biases that impacted peoples' beliefs about the state of our world. As a statistician and professor, Rosling was shocked by the number of university students (it should be noted that these are students living in Western countries) who, when presented with questions about child mortality rates, global vaccination rates, and poverty rates, tended to assume the worst. These students, enrolled in a prestigious Western university, were by and large privileged and insulated from the realities of much of the world (specifically the Global South). It turned out that these students repeatedly underestimated the progress that had been all over the world regarding these very statistics, believing that things were worse than reality

A survey circulated in 2016 revealed that 92 percent of Americans were unaware that global poverty had fallen since 1966.[41] But people in the countries affected by the fall in extreme poverty (which globally has decreased 30 percent from 1980 to today)[42] have certainly been experiencing this change in their communities, able to taste, see, and touch the progress manifesting in their lives.

Rosling calls the above tendency the gap instinct, but I believe that it is necessary to call out the privilege that fundamentally undergirds the creation of gaps between "others" and "us" (non–Global South countries), which ultimately leads us to the underestimation of risk and problems in our own communities.

This privilege gap limits our adoption of climate optimism because it builds the assumption that the people in the Global South getting hit the hardest with the consequences of climate change are quietly standing by as they watch the tides of injustice overtake their homes rather than standing against them. These factors also falsely attest that the climate crisis is only hitting people in the Global South, despite the fact that hundreds of millions of people in wealthy countries are fighting environmental struggles daily.

This gap is made obvious when examining how many privileged people are experiencing climate change as their first threat to existence. As Mary Annaïse Heglar points out, racism has been a persistent existential threat to various BIPOC (Black, Indigenous, People of Color) groups throughout history.[43] When this threat is compounded with the threat of climate change, it is no shock that BIPOC are most affected by the consequences of climate change.

The privilege gap feeds into climate doomism by perpetuating the belief that the world has never seen such a pernicious threat as climate change and that impacted communities are helplessly surrendering to the whims of climate change. In reality, many BIPOC communities have for centuries had to deal with structural threats to their existence, now being amplified by climate change. The original pernicious threat—extractive capitalism—has been assaulting marginalized folks since its inception.

The privileges many of us in the Global North are afforded can blind us from conjuring up innovative solutions to the climate crises in our neighborhoods. Many climate solutions that receive praise (and funding) in these countries have been tailored toward climate tech, capital-heavy solutions that require advanced degrees and technical knowledge to pursue. Endless solutions lie outside the bounds of a computer system or a mathematical formula; it just takes relentless imagination to find them.

OVERCOMING THE PRIVILEGE GAP

Privilege has a way of putting blinders on us, preventing us from seeing and appreciating things for what they are. Here are some ways you can challenge the privilege gap and in doing so, better appreciate the solutions to the climate crisis being pioneered by people all over the world.

1. **Follow MAPA activists.**
    ◊ MAPA is a term that stands for Most Affected People and Areas. The term is often used in parallel with the usage of the Global South. Following activists from places you have never been to and can only imagine helps to open your eyes to the incredible groundwork they are doing.

2. **Unpack your biases.**
    ◊ Easier said than done, I know. But unpacking your biases about your view of countries in the Global South and how you speak about them is essential to overcoming the privilege gap. To overcome these biases, which are typically rooted in what we

have been taught as young children, you must first address them.

3. **Acknowledge your privileges.**
   ◊ Write down things that you have that protect you from the climate crisis. Something as simple as an air conditioning unit, which many people take for granted on a scorching heat day, is a luxury for many. That water filter that you use to fill your Hydroflask with refreshing water? Also a luxury. Remind yourself of these environmental privileges you enjoy and how they might influence your experience and perception of the climate crisis.

4. **Learn about injustices in your community.**
   ◊ Injustices exist all over the world, even in the wealthiest zip codes. Learning about injustices in your community can motivate you to take action at home and open your eyes to the need for change all over.

CHAPTER 4

FAILURE OF IMAGINATION

—

"IMAGINATION IS THE ONLY WEAPON IN THE WAR AGAINST REALITY."

—LEWIS CARROLL

I recall riding home from an Easter Day brunch, full from the kids' menu sugar-dusted pancakes and impatient to get home so I could use the bathroom, despite having plenty of opportunity to do so at the restaurant before leaving. More than that, I recall my mom picking up the phone in the car and her face twisting into a grimace of pain and shock.

It is rare for a child to imagine death. But at that moment, when my mom was notified that my uncle, aunt, cousin, and grandparents had gotten in a tragic accident and that my grandpa had gone into cardiac arrest, I was confronted with the grim reality. And despite my dad comforting me like any good parent would, promising me that my nana (grandfather) would be okay, I knew he was wrong.

In the hour-long minutes that it took for us to arrive home, I braced myself for the likely possibility that my grandfather was gone. Although the next few months were brutally tough, my healing began when I imagined a life without him.

Fail to envision, envision to fail.

The failure of imagination phenomenon first materialized in the 9/11 Commission Report,[44] organized by the National Commission on Terrorist Attacks Upon the US, which charged the US Intelligence Community with a failure of imagination for not foreseeing the potential of airlines being used as weapons. In other words, the failure to prepare for the unexpected was identified as a key weakness that made us vulnerable to such heinous attacks.

This phenomenon has been especially clear during the COVID-19 pandemic. Although the world has encountered a fair share of pandemics in the past, the emergence of SARS-COVID-19 was unexpected and unprepared for. I recall being on a spring break service trip and getting a notification that the San Francisco service site had been canceled last minute due to the Coronavirus outbreaks in the area.

Our group (save for one astute and paranoid member) felt sorry for the California group but felt safe in Houston, TX. We thought it was just "one of those California things." Less than two weeks later, our university sent us an email notifying us that we would all be sent home for the semester due to the ongoing threats of the Coronavirus. We were shocked, and far from the only ones to feel this way.

The global response to COVID-19 was marred by the failure of government officials to conceptualize a worldwide pandemic that would kill millions, decimate markets, and disrupt life for four years. When we do not recognize these patterns, we fall victim to similar situations in the future. Such is the case for the already-present climate crisis.

Oftentimes, climate reports cite consequences that we will be experiencing in the next five, ten, or fifteen years if we fail to act right now. But these consequences are difficult to wrap our heads around, for both politicians and citizens alike. As a twenty-something-year-old, I find it hard to think about what I will be doing and where I will be in the next year, let alone to conceptualize what will happen in the next few decades.

Projections seem less imminent due to the lack of scientific understanding in government bodies and our myopic mental timelines that struggle to see more than two years out. Many still doubt the future threat of a two degrees Celsius increase in temperature, shrugging it off as "seeming rather small" and "not urgent."

It is all too common for people to dismiss these consequences of climate change and instead focus on the short-term benefits that might come with delaying climate action.

Failure of imagination, then, minimizes the threats that climate change poses to society while stifling the potential solutions in both the mitigative and adaptive spheres that could be used to avert the intensifying consequences of the climate crisis. It's not just about apocalyptic images of the end of times. Our failure of imagination is rooted in a failure to imagine solutions, the scalar impact of climate change, and the political/social implications of the worsening crisis.

For example, climate documentaries and commenters claiming deadly famines and resource wars are inevitable only make it more difficult to imagine the steps humans can take to mitigate such temporally distant catastrophes. In the same way negativity paralyzes change by making us feel like destruction is inevitable, so too does failure of the imagination render us unable to imagine the changes that would have to take place before we reached the extinction of a keystone species or an uninhabitable world.

The bounds of our imaginations are further limited when we surround ourselves with those who look and sound no different than us. The next chapter will expand on the importance of surrounding ourselves with those who think, look, and act differently than us, and how failing to do so impacts our ability to be optimists.

OVERCOMING FAILURE OF IMAGINATION

1. **Practice envisioning what a just climate for all would look like.**
    - ◊ It can be easy to forget what we are fighting for. Journal prompts that encourage us to envision what a healthy, equitable planet would look like for humanity are often used to help combat eco-anxiety. These prompts typically ask you to reflect on the world you are fighting for. Doing so is a helpful exercise that can encourage you to keep fighting for this utopia.

2. **Tell people what our future will look like if we fail to take action now on the climate crisis.**
    - ◊ Make the climate crisis a personal issue for people by telling them how climate change is projected to impact their community, profession, way of life, and

family. Many people are not urgently pursuing climate solutions because they do not believe it is a personal threat. Suffering from the failure of imagination, they might need to hear just how grim a future with unchecked planetary warming will be, and how they can play a role in preventing that from happening.

3. **Read about the solutions.**
    ◊ Project Drawdown is one of my favorite organizations. They are known for extensive research into the top hundred solutions to the climate crisis, gathered by experts in various fields and rated on scalability, cost, feasibility, and impact. Read the work of organizations like Drawdown that have tirelessly compiled data to encourage policymakers to invest in and promote the most impactful solutions to the climate crisis.

CHAPTER 5

ECHO CHAMBERS

—

"TYRANNY SETS UP ITS OWN ECHO CHAMBER."

—BRUCE CHATWIN

After graduating from Vanderbilt University in December of 2021, I moved back to Houston, TX. The atmosphere at my school was like that of other prestigious liberal art institutions. Most people I engaged with were quite passionate about environmental issues. Given that I was a student in the Environmental Sociology and Earth and Environmental Science programs, I never heard anyone express any distrust in climate science.

An environment of climate awareness seemed normal to me. When I came back to Houston, I gave no thought to the possibility of a changed environment. I had existed in an echo chamber for so long that I caught myself assuming that people agreed with me. So you can imagine my shock when I was chatting with someone about what I did for work and found out that they did not believe in the very phenomenon that I am dedicating my career to fighting—climate change.

Echo chambers are bubbles that perpetuate your already-held opinions. They exist on the left, the right, the middle, the outskirts, and everywhere in between. They can be empowering, validating one's belief systems and opinions. Oftentimes, they cultivate a culture of groupthink, where nuances and individual perspectives are silenced in favor of cohesive group beliefs. Had I broken out of the echo chambers I was in during college, I would have realized that many people outside of them do not believe in climate change and the threat it poses to humanity.

Social media algorithms propagate these echo chambers, showing users more of what they agree with and less of what they do not know or agree with.[45] Because I follow many other climate activists and educators, my timelines are often filled with frightening studies about the climate crisis and attention-grabbing tweets that sometimes push me to the point of panic attacks.

I have found myself in echo chambers that perpetuate climate doomism, the nihilistic worldview reflecting the belief in an inevitable demise of our planet. There are countless videos with millions of likes where twenty-something-year-olds lament that the "downfall of civilization [is] happening during my twenties" and that "it is too late to make a difference." It is important to note that these people are often not scientists and are not making any science-backed claims. Instead, they are projecting their feelings about the negative news that circulates and proliferates around them, which is a valid response to have.

These projections can be immensely harmful, pushing people into algorithms that perseverate on this nihilism, eventually convincing them that acting now is pointless and fruitless—the belief that is a most harmful enemy to the climate action our planet so desperately needs.

In a study on the spread of information related to dreadful news involving seemingly uncontrollable phenomena, researchers found that as information was passed along chains, messaging became increasingly negative, despite reexposure to the initial information.[46] Given that, as previously noted, disaster framing is most frequently used in headlines and the first few lines of an article, I can only imagine how climate news stories get perceived as increasingly dire as they are shared from person to person within an echo chamber.

Stepping outside of our echo chambers allows us to break free from doomism rhetoric and have a higher potential for impact by interacting with those who likely aren't as fired up for the planet as you are (we'll talk about that later).

Convincing a person who previously did not prioritize climate change of its importance will most likely produce a domino effect with impact accelerating, compared to engaging with someone who already acts with climate change first in mind. In this way, breaking free of echo chambers can embolden our climate optimism by allowing us to engage in personal and high-impact change.

BREAKING FREE OF ECHO CHAMBERS

1. **Make it a point to engage with opinions you do not fully agree with.**
   ◊ Although this can be a sometimes-frustrating exercise, it is so important to learn from those you disagree with through engaging with them.

2. **Make room for all in the climate movement.**
    ◊ Far too often, I see liberal climate activists shunning those with different political viewpoints from entering climate spaces. This practice can be harmful, preventing us from building bridges and moving the needle on climate action. We must include diverse perspectives and voices representative of our world in climate spaces. If we fail to do so, we are deluding ourselves with a false image of our world.

TAKEAWAYS FROM PART I

—

All these factors and biases—information overload, negativity instinct, privilege gaps, failure of imagination, and echo chambers—make it undeniably challenging for someone like you and me to cultivate a mindset of optimism. Climate optimism becomes even more difficult to cultivate considering the sordid projections of humanity's future.

They also make it more important for us to consciously seek out the positive—in all the progress the world has already made and all that we have left to make. We must do so because it is the only way to move forward. Hope propels us forward. Without it, we are stuck in cycles of environmental degradation—and worse—doomed to irreversible damage.

Though our window to act is narrowing, our portal of opportunities is expanding.

This mantra guides my work and everything that follows in the book. I have held onto this mantra throughout the last few years of my activism journey. These few words have pulled me out from depths of despair and days where all I find myself able to do is scroll through TikTok, soaking in the doomism.

I recognize that it is not a switch you can flip. This mindset can take work to cultivate. My job is to gently guide you toward the framework of climate optimism and hopefully convince you of why it is necessary and informed.

Climate optimism should be informed, but not conditional.

In the next section, I'll unpack some of the most inspiring places in which progress has been made and root out the popular misconception that we have done nothing for our planet in the last decade by documenting examples of immense progress in a variety of fields and sectors.

Hope is what will carry us through. Without it, there is no chance for progress.

POEM BY MORIAH LAVEY

a look into my crystal ball

> "We are dancing flames committed to conquering darkness and to challenging those who threaten the planet and the magic and mystery of life"
>
> In the future I see pain,
> Earth shattering loss,
> irrevocable change,
> lives, species, infrastructure gone.
>
> As oceans surge,
> storms gain strength with less protection,
> bringing unpredictable destruction in their wakes.

As temperatures fluctuate,
natural balance thrown off,
mass migration disrupted by our ignorance.

As disease will thrive,
innocent will die.
What can we do?

I am intent on exploring,
loving, cherishing
Our Mother in all She is.
giving Her the love and light She deserves
following centuries of neglect.

I am intent on showing others the endless potential in this
world and unifying,
dispelling darkness with lovingkindness,
destroying imaginary lines marking imaginary places.

Creating communities built on compassion
not oppression and selfishness.
laws governed by Nature,
not by those in Forbes, in touch with greed
and macroeconomics,
out of touch with reality.

Those enlightened have been awoken
to the wrongness of society
to the lives stolen and lost
onto whose backs we have built this Dream.

and we will not rest until the paradigm is shifted.

In the future I see hope,
humanity rising to the challenge.
Our house is literally aflame,
interconnection and acceptance must prevail.

Do not fear, act.
Another world is not only possible,
She is on the way.
on a quiet day,
I can hear her breathing.

PART II

—

LOOKING BACK TO LOOK FORWARD

CHAPTER 6

THINGS AREN'T ALWAYS AS THEY SEEM

—

"MEMORY PRODUCES HOPE IN THE SAME WAY THAT AMNESIA PRODUCES DESPAIR."

—WALTER BRUEGGEMANN

In the previous section, we explored how, because of our many biases and the increased interconnectivity of the world, things seem bleak.

The world seems like it is worse off than ever before; at best, stagnating in inequality and disasters and at worst, declining in rights and health. The keyword here is *seems*.

Here is my take: increased global connectivity harnessed by social media, in conjunction with the various cognitive biases inflamed by media consumption, has enhanced our interpretation of conflicts.

By many measures, humanity is better off than ever before, but climate change is disrupting this progress by amplifying conflicts to a degree we have not yet witnessed.[47]

Scientists know what the problem is and how to fix it. They even have all the technology we need to mitigate the climate crisis. Many citizens know about the scope of the problem and its consequences, even realizing the need for systemic change, but struggle with being a part of that change and convincing others of its importance. The greatest challenge in conquering the crisis lies in our ability to deploy the technology in a fast and fair manner and get citizens involved in ways that feel promising for them.

Our conversations around climate change, influenced by media rhetoric, are too confined to the scope of the problems rather than the scope of the solutions.

As such, we are blind to the stories of environmental progress that happen every day across all parts of the globe, while hyper-focused on the threats of what will happen if we don't take action. No wonder so many of us have eco-anxiety.

Despite our insufficient progress in successfully shifting the collective will of the Western world, which has had an outsized impact on the causes of climate change, there is some progress we *can* celebrate to encourage and remind us that change has happened before and can happen again.

These boulders of change—driven by concerned citizens like you and me—can cascade, instigating global change and toppling unjust systems of inequality that have been operative in society for far too long.

This section explores these trends and unpacks how they can be leveraged to build a better world. I will restrain myself from talking about the climate innovations across every sector that provide us with auspicious projections for decarbonization, as there are already standout books on the topic. The encouraging trends that I have picked up on won't be measured by their greenhouse gas reduction potential or economic costs, but rather the power they have to propel this boulder effect, tearing down mindsets and systems of oppression that have fueled environmental degradation.

Before getting into those trends, let's revisit a few oft-forgotten examples of scientists expressing environmental concerns and government officials listening and putting forth effective policies. Don't believe me? Check this out.

1. **Sulfur dioxide was reduced by 80 percent in the USA after implementing a cap-and-trade acid rain program.**[48]
    Due to international concern regarding the accumulation of ecosystem-destructing sulfur dioxide in the atmosphere, most of which was generated from coal-powered plants, Congress included an SO2 allowance trading program within the Clean Air Amendments of 1990.[49]

    The program allowed sources to have freedom in selecting their method of reducing these emissions. This system resulted in emissions declining at a rate much faster than that which was anticipated, for a quarter of the cost.[50]

    Had this policy not been so successful, the high levels of atmospheric sulfur dioxide and resultant acid rain would have resulted in a significant deterioration of soil quality and contamination of food systems.

2. **A ban on refrigerants called CFCs (chlorofluorocarbons) led to stratospheric ozone rehealing.**
In the late '80s, scientists urged policymakers to consider the CFCs and other ozone-depleting chemicals chipping away at our ozone layer, exposing us to UV rays and making us more vulnerable to global warming. Global leaders listened, and the Montreal Protocol became the first UN treaty to be ratified by every member state. As a result of the protocol's call for countries to halt the production and consumption of ozone-depleting chemicals, the upper stratospheric ozone has increased 1 to 3 percent per decade since 2000 and it is estimated that by the 2030s, northern hemisphere mid-latitude total column ozone will return to 1980 levels.[51]

Scientists estimate that without actions spurred by the Montreal Protocol, there could have been 325–690 billion tons less carbon held in plants and soils by the end of the century...in other words, 325–690 billion more tons of carbon heating our atmosphere.[52]

3. **Clean Air Act protects our bodies.**
A landmark piece of climate legislation—the Clean Air Act—was passed in 1970 and has done more for the health of Americans than we could imagine. Since 1990, the concentrations of six pollutants have decreased by an average of 64 percent. Ambient concentrations of lead, sulfur dioxide, and carbon monoxide have declined by 98 percent, 88 percent, and 77 percent, respectively, since 1990.[53]

Because of this act, an estimated 205,000 premature deaths were averted within the first twenty years of its implementation.[54] The positive impacts of the legislation are ongoing.

Bipartisan support (collaboration between two opposing political parties) for this act pushed its progress forward. As a result, air pollution has decreased even as the US economy has grown.[55]

QUIZ TIME

I am inspired by the exercises in the book *Factfulness* that I mentioned in Part I, Chapter 3 (can you tell it is one of my favorite books?) to test your understanding of some environmental trends in the last few decades.

1. What percentage of proposed coal plants have been canceled in the past five years?[56]

    A. 72 percent   C. 34 percent

    B. 20 percent   D. 76 percent

2. What percentage of vehicles do auto manufacturers expect to be all electric in the US by 2030?[57]

    A. Over 30 percent

    B. Over 20 percent

    C. Over 75 percent

    D. Over 50 percent

3. How many local governments in the US have established Climate Action Plans?[58]

   A. 25
   B. 50
   C. 300
   D. 600

4. What is the number of Americans working in clean energy?[59]

   A. 2 million
   B. 1.4 million
   C. 3 million
   D. 1.8 million

5. How much has the cost of photovoltaic (PV) modules, a key part used in solar panels, decreased in the past four decades?[60]

   A. 22 percent
   B. 99 percent
   C. 45 percent
   D. 80 percent

6. What percentage of Americans believe that environmental protection trades off jobs and economic growth?[61]

   A. 25 percent
   B. 32 percent
   C. 17 percent
   D. 21 percent

7. Which of these animals has moved off the endangered species list?

   A. Giant Panda
   B. Gray Wolf
   C. Louisiana Black Bear
   D. Southern White Rhinoceros
   E. All of the above

Answer Key
1. D
2. D
3. D
4. C
5. B
6. C
7. E

Hopefully you too found that you underestimated some of the good things happening for our planet earth. Although the prospects for our planet are troubling, glimmers of hope in the past exist and are important to revisit to look forward. Acknowledging past wins allows us to unlock the framework of climate optimism and help us achieve wins for the future.

Speaking of past wins, now might be a great time to take a breather from the statistics and enjoy some easier-to-read headlines. Since I started my journey on climate optimism, I have been researching news stories that give us hope for the planet. This exercise has helped me seek out the good amidst the onslaught of bad news.

From the practice of collecting "Weekly Earth Wins," I have identified three spaces: the courts, the streets, and the markets, which give me the most hope for the climate crisis. We will unpack those in the next section, so buckle up.

Before that, let me offer you my favorite list of all: every Earth Win I have documented since starting the series in 2021.

CHAPTER 7

LET'S TALK ABOUT EARTH WINS

—

I look at this list whenever I feel like it's all going to sh*t, for lack of better words, and I recommend you do so too.

WINS IN THE COURTS

Oil giant Shell was declared liable in a spill case filed by Nigerian villagers of Oruma and Goi.[62]

◊ Thirteen years after two pipeline spills in the Nigerian villages of Oruma and Goi, farming communities got justice in the form of damages payment. Decided by an appeals court in The Hague, this case was in part significant because it was an instance of a multinational corporation held accountable for actions done overseas by courts at home.

Four NGOs (non-government organizations) won a court case against the French government for its failure to uphold climate commitments.[63]

◊ After successfully garnering support from over two million people, four French environmental groups successfully argued that the French State failed to uphold its commitments to combating global warming (more specifically to reduce greenhouse gas emissions by 40 percent by 2030). The court's decision effectively rendered climate inaction illegal and compensation for ecological damage admissible.

Fracking was officially banned in the Delaware River Watershed.[64]

◊ The fifteen million people who rely on the Delaware River Watershed were spared from the threat of chemicals leaching into the water after the Delaware River Basin Commission voted to permanently ban hydraulic fracturing for natural gas in the watershed, affirming a previous moratorium put in place.

Brazil's Supreme Court accepted an appeal by an Indigenous group and agreed to review the process around a past case that canceled the demarcation of the tribe's territory.[65]

◊ In response to their previous ruling that the Guarani Kaiowá people have no legal claim to their ancestral territory, the Brazilian Supreme Court unanimously decided to reopen the case and rehear the argument for demarcation with the inclusion of Indigenous groups' inputs.

New Orleans City Council ordered the city's main utility company to go 100 percent carbon free by 2050.[66]

◊ The main utility company powering New Orleans—Entergy Corp.—now has the responsibility to become 90 percent carbon free by 2040 and 100 percent carbon free by 2050. Thanks to the New Orleans City Council, Entergy Corp. pushed its goal of being net-zero by 2050 further to emitting no carbon by the same date. This timeline is one of the fastest in the Southern United States.

European Court ordered Shell to cut carbon emissions by 45 percent by 2030.[67]

◊ Along with 17,000 other co-plaintiffs, Friends of the Earth Netherlands successfully argued that Shell's sustainability policy violated the duty of care the company has to citizens and that the company must be aligned with the Paris Climate Agreement. The plaintiffs communicated that the company's knowledge of the climate crisis and its outsized role in it long before the twenty-first century violated the rights enshrined in the European Convention on Human Rights.

TC Energy Corp officially revoked the permit for the Keystone XL Pipeline.[68]

◊ After countless hours of work by grassroots activist groups and environmental/land defenders, the Keystone XL Pipeline finally has ended with the cancelation of the presidential permit for the pipeline's border crossing.

Maine became the first US state to pass a law requiring all state funds to be divested from fossil fuel assets.[69]

◊ By 2026, Maine will be required to sell off $1.3 billion worth of fossil fuel investments from its state cash pool and public employee retirement system.

Oil companies have been ordered to pay part of a $7.2 billion tab to retire hundreds of aged wells in the Gulf of Mexico.[70]

◊ Aging wells in the Gulf of Mexico need to be fixed, and a federal judge ruled that companies, including ExxonMobil Corp., BP PLC., Hess Corp., and Royal Dutch Shell PLC, would have to pay part of the $7.2 billion tab, as the owner of these old wells has filed for bankruptcy protection for the second time.

Washington Supreme Court made history as the highest court in any state to recognize the climate necessity defense.[71]

◊ The state's recognition of the climate necessity defense, which we will talk in detail about later, is significant in that it validates an informal legal instrument, potentially paving the way for environmental defenders to be absolved of criminality in obstructing property or interfering with operations of destructive companies.

The French government was fined by their Supreme Court for failure to improve air quality.[72]

◊ The highest administrative court in the state fined the country ten million euros for failing to improve levels of nitrous dioxide in the air, four years after the government was ordered to reduce levels in alignment with European living standards.

A major drilling project proposed in Alaska by ConocoPhillips has been suspended by a federal judge.[73]

◊ A US District Court Judge discarded approval for an oil project on Alaska's North Slope after assessing a flawed environmental review.

Montana tribe received $1.9 billion in a water rights settlement.[74]

◊ The Confederated Salish and Kootenai Tribes of Montana finalized a decades-in-the-making settlement confirming the tribes' water rights and control of the National Bison Range, along with funding for habitat restoration. This is the largest settlement ever awarded in an Indigenous water rights case.

British Columbia paid the Blueberry River First Nations sixty-five million dollars.[75]

◊ After the BC Supreme Court ruled that the Blueberry River First Nations "Treaty No. 8 Rights" had been violated by permitting and industrial development, British Columbia signed an agreement with the Blueberry River First Nations allocating money for land, river, wetland, and seismic line restoration.

Forty-six countries across Europe joined together to establish a legally binding mechanism that protects environmental defenders.[76]

◊ A new mechanism will be added to the United Nations to protect environmental defenders who are often the victims of unfair punishment and violence. The Aarhus Convention will ensure that defenders are not persecuted or harassed as they take on the important task of saving our planet.

New Yorkers approved a ballot measure adding the right to clean air and water to the state constitution.[77]

◊ The right to clean air will now be enshrined in the state constitution, adding pressure on the government to include environmental justice in policy decisions.

A landmark ruling blocked mining in Los Cedros, an Ecuadorian forest.[78]

◊ While Ecuador's constitution had included the Rights of Nature, judicial precedent for establishing these rights where corporate interests exist had not been established. This landmark ruling protecting the cloud forest and all its inhabitants is encouraging for the prioritization of nature over industry.

Three oil companies were charged for an October oil spill in Southern California.[79]

◊ Amplify Energy Corp., Beta Operating Co., and San Pedro Bay Pipeline Co. have been charged with criminal negligence after illegally discharging 25,000 gallons of oil into federal waters and deceiving the public about their awareness.

A federal judge invalidated the largest offshore oil and gas lease sale in the nation's history.[80]

◊ 1.7 million acres of drilling leases in the Gulf of Mexico have been canceled due to a faulty environmental assessment.

Ecuador's highest court ruled that Indigenous communities must be consulted and give consent for extractive projects on or near their territory.[81]

◊ Free, prior, and informed consent from Indigenous communities is now constitutionally required for any extractive projects operating on or near the territory. After centuries of Indigenous erasure and negligence, the concept of free, prior, and informed consent is essential for giving Indigenous communities the rights they were stripped of and our planet the protection Indigenous communities have provided it.

Five student-led divestment campaigns filed legal complaints, alleging that their university's investment in fossil fuels violates fiduciary responsibilities outlined in UPMIFA.[82]

◊ Using a relatively nascent legal strategy, a student group (that I was part of) filed legal complaints against universities to challenge the legal standing of investing in extractive companies and industries.

An appeals court lifted a ban that blocked the federal government from considering "the social cost of carbon."[83]

◊ The social cost of carbon, which calculates the societal damage (in dollars) associated with each ton of carbon dioxide, is instrumental to just policy-making decisions regarding environmental action. After a district court in Louisiana had banned federal agencies from using this number in carbon accounting, an appeals court judge stayed the order.

Hawaii circuit court judge ruled in favor of Honolulu against Big Oil companies' attempts to dismiss the city's lawsuit holding

these companies accountable for using deception tactics to mislead the public about their role in climate change.[84]

◊ The attempts of Chevron, Sunoco, and ExxonMobil to dismiss a lawsuit from the Honolulu Board of Water Supply accusing the companies of engaging in deception tactics to discredit climate change consensus faltered after a circuit court judge ruled in favor of the city. The decision was regarded as unprecedented and immensely important for protecting community rights over big corporations.

UK's Net-Zero Strategy found to breach Climate Change Act, holding lawmakers and politicians accountable for approving accurate carbon budgets.[85]

◊ In a historic ruling, a UK court found that the country's Net-Zero Strategy does not fulfill obligations made by the government under the Climate Change Act. In this important act of accountability, the judiciary acts as a check to ensure that the legislative branch is making good on its environmental commitments.

WINS IN THE STREETS

For the first time, the US Department of Transportation made a budget for climate justice projects.[86]

◊ Under the leadership of Secretary Pete Buttigieg, the US Department of Transportation announced that the Infrastructure for Rebuilding America program would include funding for environmental justice and climate change projects, emphasizing the importance of including climate justice in transportation policy.

Jersey City, USA, announced plans to open vertical farming in two public housing locations.[87]

◊ A new partnership between vertical farming company AeroFarms and the Jersey City Housing Authority allows low-income families to access nutritious food grown efficiently and sustainably.

Michael Regan became the first Black man to head the EPA.[88]

◊ Coming from his position as the former secretary of North Carolina's Department of Environmental Quality, Regan received bipartisan support for his appointment to the head of the EPA.

The United Kingdom announced Right to Repair legislation for certain appliances.[89]

◊ Several types of appliances sold in EU countries must be made repairable for up to ten years with the passage of the Right to Repair legislation. This legislation challenges the common phenomenon of planned obsolescence, where technology or appliance companies manufacture products that cannot be repaired, thus fueling an unsustainable waste cycle.

South Australia became the first jurisdiction in Australia to ban single-use plastics.[90]

◊ Single-use plastic cutlery, stirrers, and food and beverage containers will now be prohibited from sale, supply, and distribution under the Plastic Reduction Act 2021.

Brenda Mallory was confirmed as the head of the White House Council on Environmental Quality and will be the first Black person to serve in the role.[91]

◊ Environmental lawyer Brenda Mallory makes history as the first Black person to serve as the CEQ chair, where she will prioritize rectifying environmental injustices in the management of energy and environmental policy.

The state of Massachusetts passed landmark bipartisan climate change legislation that puts the state on track to reach carbon neutrality by 2050.[92]

◊ Governor Charlie Baker signed legislation codifying the commitment to achieve net-zero emissions by 2050 while increasing protections for environmental justice communities.

Governor Jay Inslee of Washington signed the Climate Commitment Act.[93]

◊ The Climate Commitment Act funds investments in air quality improvements for overburdened communities, transit access and mobility equity, electrification, and implements carbon caps. The Act passed 54-44 in the Washington House of Representatives.

San Diego County created an office of Climate and Environmental Justice.[94]

◊ The office will reduce pollution's impact on the county's poorest residents and help county officials incorporate environmental justice into policy and decision-making.

Chile became the first Latin American country to ban single-use plastics in food sales and delivery.[95]

◊ At least 23,240 tons of single-use plastics are generated each year by food delivery in Chile, a mere 4 percent of which is recycled. This initiative will help to cut the number of plastic wastes created by banning single-use plastics in food distribution.

Turkey banned plastic waste imports.[96]

◊ After an investigation revealing British plastic waste on Turkish beaches, the Turkish government announced a ban on imports of most types of plastic waste.

Three oil companies relinquished leases in the Arctic.[97]

◊ After a formal order from Department of Interior Secretary Deb Haaland halting drilling leases in the Arctic Refuge, political battles, and various economic factors, three major oil companies give up on drilling leases to the Arctic National Wildlife Refuge.

The federal government transferred eighty acres of land on O'ahu to Native Hawaiians for homesteads.[98]

◊ Eighty acres of surplus federal property have officially been transferred to Native Hawaiians. As part of the Hawaiian Home Lands Trust, this land will provide homesteads for around two to four hundred Native Hawaiian families.

China and America signed onto the Kigali Amendment of the Montreal Protocol, which commits the country to the phasing out of dangerous chemicals.[99]

◊ The Kigali Amendment is an important piece of global legislation aimed at phasing out ozone-destroying chemicals called hydrofluorocarbons (HFCs). If fully implemented, the amendment can prevent more than 0.4 degrees Celsius of warming by the end of the century.

The Environmental Protection Agency announced a fifty-million-dollar project to fund environmental justice.[100]

◊ Under the American Rescue Plan, the EPA has allocated fifty million dollars for environmental justice grants and to help low-income communities of color get jobs in the environmental sector. This money would support the YH2O mentoring program in Baltimore City, which prepares young adults to be employed in full-time jobs within the water industry.

Culver City, CA, began plans to phase out oil drilling in what is the largest urban oil field in the United States.[101]

◊ The Culver City Council voted four to one in favor of an ordinance that would phase out oil production and require the cleanup of the well site within five years.

The Department of Interior returned 18,800 acres of land to Montana Indigenous Tribes.[102]

◊ Land known as the National Bison Range was officially returned to the Confederated Salish and Kootenai Tribes of the Flathead Reservation in Montana.

Due to community organizing, plans to build the Byhalia connection pipeline through Memphis were officially canceled.[103]

◊ A pipeline proposed by Valero and Plains All American Pipeline for construction through South Memphis, a predominantly Black community, was officially nixed due to community organizing.

Argentina became the first country to ban open-net salmon farming due to environmental impact.[104]

◊ Open-net fish farming techniques can have dangerous consequences, like sea lice infections, algae blooms, and toxic water conditions. The move from the Argentinian government is a win for animal and environmental rights.

Greenland committed to stopping all exploration for new oil and gas reserves.[105]

◊ Emphasizing the country's commitment to the green transition, Greenland's natural resources minister committed to halting all further oil and gas reserve exploration.

Maine passed a law to make big companies pay for the recycling of their packaging.[106]

◊ The first extended producer responsibility law in the US was passed with Maine's recycling reform law, which forces corporations to take responsibility for the cost of handling hard-to-recycle packaging.

Karachi's largest public hospital announced its commitment to run on 100 percent renewable energy.[107]

◊ With assistance from the World Bank, the largest public sector hospital in Karachi, Pakistan, is building a solar system to power 100 percent of hospital operations with clean energy.

The Canadian government announced plans to give almost eight billion dollars to the First Nations to fix water quality problems.[108]

◊ After a class action lawsuit over the lack of safe drinking water in First Nations communities, an agreement was reached to provide the First Nations with nearly eight billion dollars to fix water quality problems.

Norway dedicated over a billion dollars to green energy investments in nations in the Global South.[109]

◊ The Norwegian Investment Fund administered a one-billion-dollar climate fund with the purpose of reducing developing nations' dependency on fossil fuels.

A county in Washington became the first in the US to ban new fossil fuel infrastructure.[110]

◊ The county of Whatcom, Washington, is the first jurisdiction in the US to ban new fossil fuel infrastructure after a unanimous vote by the county council.

New York state granted over fifty million dollars in funding to a program that will increase access to community solar, resulting in electric bill savings for 50,000 low-to-moderate income households.[111]

◊ In efforts to progress a just energy transition, the New York governor launched the Inclusive Community Solar Adder program, which makes fifty million dollars available for community solar projects supporting disadvantaged New York communities.

A company has agreed to cut two proposed dam projects on the Colorado River after protests from Indigenous communities in the area.[112]

◊ Two of the three proposals to build hydroelectric dams in the Little Colorado River watershed were withdrawn from the developer following pushback from the Navajo, Zuni, and Hopi people.

The first US federal office aimed at understanding the connection between climate change and human health was established.[113]

◊ The US Department of Health and Human Services established the Office of Climate Change and Health Equity to protect vulnerable communities that bear the brunt of pollution and climate disasters at the expense of public health.

The world's last toxic leaded gasoline refinery shut down.[114]

◊ Algeria finished up the world's last stockpile of leaded gasoline, ushering in a turn away from the usage of the toxic fuel.

Ebony Twilley Martin became the first Black woman executive director of a national legacy environmental organization in the US.[115]

◊ The first Black executive director of Greenpeace USA, Twilley Martin plans to bring environmental justice and racial disparities to the forefront of climate work.

The Los Angeles City Council voted unanimously to end new oil and gas drilling.[116]

◊ The city council voted to ban new oil and gas wells and phase out existing wells over a period of five years, a huge win for environmental justice in a city with one of the largest urban oil fields in the country.

England became the first country in the world to require all new homes and offices to have electric vehicle chargers.[117]

◊ With this legislation, all new homes and offices in England will be required to have smart charging devices that can charge cars during off-peak periods.

Four national parks in Australia—adding up to over 988,400 acres of land—were handed back to their Aboriginal owners, the Eastern Kuku Yalanji people.[118]

◊ The original Eastern Kuku Yalanji people, believed to inhabit the area for over 50,000 years, will now own and comanage the UNESCO World Heritage Site Daintree, Ngalba-bulal,

Kalkajaka, and the Hope Islands National Parks, with a new nature refuge also slated to be established.

Philadelphia implemented a ban on single-use plastic bags.[119]
- ◊ After years of failed attempts, the city of Philadelphia passed a ban on single-use plastic bags.

PennEast gas pipeline suspended construction after protests.[120]
- ◊ The natural gas pipeline, stretching from Pennsylvania to New Jersey, has officially ceased further development due to regulatory hurdles and community protests.

The DC area Audubon Society announced that it would divorce itself from its namesake, John James Audubon, who was a slave owner.[121]
- ◊ After 124 years of association with anti-abolitionist John James Audubon, the oldest independent environmental organization in the DC area detached itself from the organization's namesake, citing the racist beliefs of John James Audubon that are in misalignment with those of the organization.

England passed a long-awaited act that provides a framework for legally binding environmental targets in four categories.[122]
- ◊ The UK government passed the Environment Act, which includes a legally binding 2030 species abundance target and establishes an environmental watchdog body to

hold the government accountable for compliance with environmental laws.

For the first time, a Native American person was confirmed to be the head of the US National Park Service.[123]

◊ In 2021 the Senate unanimously approved the nomination of Charles "Chuck" Sams III to lead the National Park Service, making him the first Native American to lead in that position.

Justice40 initiative was launched.[124]

◊ A win for environmental justice, the Justice40 initiative created by the Biden administration sets a goal of 40 percent of all federal investments directly benefiting disadvantaged communities that environmental harms have burdened.

After over a year of protesting, Indian farmers saw a victory with the prime minister's reversal of agricultural laws.[125]

◊ The "farm laws" enacted by the right-wing Bharatiya Janata Party (BJP) to deregulate industry and considered by farmers to be anti-small farmer were officially repealed.

523 acres of CA redwood forest were handed over to ten Native American tribes and designated as tribal protected areas.[126]

◊ Save the Redwoods League, a Californian conservation group, purchased over five hundred acres in Mendocino

County to donate and transfer to a consortium of ten Northern Californian tribal nations.

Honduras banned open-pit mining.[127]

◊ Honduran President Xiomara Castro banned the surface mining technique of extracting minerals from an open pit in the ground, also known as open-pit mining, for its harmful effects in exposing waste to the environment.

The governor of California proposed a plan that would give a hundred million dollars to Indigenous leaders to buy ancestral lands.[128]

◊ As part of the state's plan to preserve one-third of land and coastal waters by 2030, the California governor stressed the importance of having Indigenous leaders and communities at the forefront of preservation decisions and policies, proposing that the allocation of the hundred million dollars would be determined by these frontline communities.

The Rappahannock Tribe received 465 acres of land back in Virginia, USA.[129]

◊ Over three hundred years after the Rappahannock Tribe was forcibly removed from Native lands on the Chesapeake Bay, some of the land has been returned to them due to tribal efforts and the work of the Chesapeake Conservancy.

New poll data showed that a majority of Americans value environmental protection over economic growth.[130]

◊ The traditional notion that Americans value economic prosperity over environmental protection is, in fact, not accurate, as proven by recent Gallup poll data showing that despite economic concerns trumping environmental concerns during the 2008 financial crisis, concern for the environment has once again superseded that for the economy.

Québec became the first jurisdiction in the world to explicitly ban oil and gas development in its territory.[131]

◊ Québec's National Assembly voted to ban all new oil and gas exploration and shut down existing drill sites and wells (around a thousand of which exist) within three years.

Tasmania became a carbon-negative country.[132]

◊ After efforts to reduce logging, Tasmania has joined the handful of countries that have gone from being a carbon-neutral to a carbon-negative country, meaning they remove more carbon dioxide from the atmosphere than they emit.

London signed the Fossil Fuel Non-Proliferation Treaty.[133]

◊ London became the largest city to join the Fossil Fuel Non-Proliferation Treaty, which calls for phasing out fossil fuels and implementing a just energy transition.

The Plastic Pollution Producer Responsibility Act passed in California, mandating that companies cut down plastic packaging by 25 percent within the next ten years and that the state increase plastic recycling rates by 63 percent.[134]

- ◊ Considered the strongest plastic reduction policy in America to date, this bill passed through the California Legislature with little opposition. According to an analysis by the Ocean Conservancy, the bill could prevent almost twenty-three million tons of plastic waste from being generated in the next ten years.

Massachusetts moved on a bill that would allow residents to contribute to climate-vulnerable countries while filing their tax returns.[135]

- ◊ If passed (very likely), Massachusetts would become the first subnational authority to allow individuals to contribute to the United Nations Least Developed Countries Fund (LDCF), which helps the most climate-vulnerable countries develop resiliency from climate change.

Hawaii closed its last coal power plant.[136]

- ◊ The last Hawaiian coal plant in Oahu, which supplied around 11 percent of the state's electricity in 2021, officially closed as part of the state's initiatives of switching to 100 percent renewables by 2045.

Train line in Germany switched to hydrogen-powered transportation.[137]

◊ The state of Lower Saxony rolled out a fleet of hydrogen-powered passenger trains, intending to replace all diesel-powered trains with this zero-emissions alternative in the next year.

WINS IN THE MARKETS

Solar projected to become the cheapest electricity in the US by 2030.[138]

◊ In a nod to the decreasing costs of solar and increasing environmental and capital costs of fossil fuels, a report by Wood Mackenzie projected that solar will become the cheapest electricity in the US.

IKEA committed to saving 11,000 acres of US forest from development.[139]

◊ The Swedish furniture company acquired 11,000 acres of land in southeast Georgia from The Conservation Fund to protect it from development.

Citigroup became the first American bank to refuse new clients that have plans for new or newly expanded coal plants.[140]

◊ The American bank published an updated Environmental and Social Policy Framework, which declares a commitment not to take any new clients with plans to expand coal-fired generation.

A retired German coal plant announced plans to convert to a Green Hydrogen fuel plant.[141]

◊ The country's most prominent coal power plant is decommissioned and a letter of intent for building a hydrogen plant was signed.

Toy company Mattel launched a take-back program called "Mattel PlayBack" designed to recover and reuse materials in old toys for future Mattel products.[142]

◊ The program aligns with the company-wide goal of achieving 100 percent recycled, recyclable, or bio-based plastic across all products and packaging by the year 2030.

Mercedes-Benz invested in a Swedish carbon-free steel startup called H2 Green Steel and plans to launch green steel in their vehicles by 2025.[143]

◊ Steel production and manufacturing is an incredibly carbon-intensive process, contributing an estimated quarter of industry-related carbon dioxide emissions. Mercedes-Benz's investment in carbon-free steel and plan to include it in vehicles is a significant step for the industry.

The richest university on earth has committed to divesting from fossil fuels.[144]

◊ After years of student-led activism, Harvard University in the US finally committed to divesting its fifty-three-billion-dollar endowment from fossil fuel companies, setting a precedent for other universities around the world.

French luxury conglomerate Kering has banned the use of fur.[145]

◊ With brands like Gucci, Yves Saint Laurent, Balenciaga, and Alexander McQueen under their belt, banning all animal furs across brands by fall of 2022 is a significant win for animal rights.

The MacArthur Foundation, an eight-billion-dollar foundation, committed to divesting from fossil fuels.[146]

◊ With this divestment announcement, the MacArthur Foundation became the largest foundation (of yet) to move money away from the extractive oil and gas industry.

Google and YouTube announced a plan to remove ad funding for climate deniers.[147]

◊ In light of the proliferation of misinformation and disinformation on online platforms and the resultant consequences, Google and YouTube will no longer allow creators who deny the existence of climate change to monetize their content.

University of Toronto announced plans to divest from fossil fuels.[148]

◊ University president Meric Gertler announced a divestment from all direct investments in fossil fuel companies and an eventual divestment from indirect investments by 2030.

Apple announces the introduction of a Right to Repair program.[149]

◊ After receiving heat for the planned obsolescence of old Apple products and the Federal Trade Commission's vote to draft new policies requiring manufacturers to provide strict repair restrictions, Apple announced the inclusion of parts, tools, and instructions for consumers to more easily repair their devices.

United Airlines flies the first full passenger flight with 100 percent sustainable fuel.[150]

◊ The United Airlines Boeing 737 MAX 8 flight was one of the first flights to use 100 percent sustainable aviation fuel (SAF) for an engine, as current regulations limit a 50-50 mix of regular jet fuel and SAFs. This fuel emits 80 percent less CO_2 than traditional jet fuel.

Dartmouth University divested from fossil fuels.[151]

◊ Citing promising returns from sustainable energy investments and the imperative to act on climate change, Dartmouth University announced the divestment of the endowment from fossil fuels and an investment of four hundred million dollars into programs and institutes related to solving the climate crisis.

The world's largest green hydrogen hub announced a Texas opening in 2026.[152]

◊ Green hydrogen is a transformative zero-emissions energy source that has the potential to transform the way we power our world. Hydrogen City, as it's called, will open in south Texas in 2026 to store up to six TWh of energy.

Starbucks CEO committed to phasing out paper cups and allowing customers to use a provided or personal to-go mug by the end of 2025.[153]

◊ As the largest coffee company in the world, Starbucks announced that part of its waste-reduction plan would entail providing customers with branded reusables.

The US Securities and Exchange Commission (SEC) proposed a new rule requiring all publicly traded companies to disclose their greenhouse gas emissions and climate risks.[154]

◊ The SEC, a regulatory body created to protect investors, maintain fair, orderly, and efficient markets, and facilitate capital formation, proposed new disclosure rules that would overhaul the relaxed ESG reporting requirements.

Boston University has committed to divesting from fossil fuels.[155]

◊ The university board of trustees voted in favor of adopting a plan to move investments tied to the fossil fuel industry.

A coalition of tech companies announced plans to purchase $925 million worth of permanent carbon removal from companies developing the technology over the next nine years.[156]

◊ Led by Stripe, these companies joined together to help fund carbon capture solutions in light of the Intergovernmental Panel on Climate Change (IPCC) estimates that we must remove an average of six billion tons of carbon dioxide from the atmosphere by 2050 to limit warming by 1.5 degrees Celsius.

In the annual shareholder meetings for Citi, Wells Fargo, Bank of America, and Goldman Sachs, resolutions to stop the banks from investing in fossil fuel expansion received record shareholder votes.[157]

◊ Despite the shareholder votes falling significantly short of what was needed to pass the resolution, it is quite significant that at least 10 percent of shareholders voted in favor of divestment, signifying a wave of change in banking.

Near 100 percent of electricity demand was powered by renewables for the entire state of California for the first time.[158]

◊ 99.87 percent of the state's energy was served by renewables, breaking previous records.

MORE MISCELLANEOUS WINS

MIT engineers designed a conceptual hybrid airplane that could eliminate 95 percent of toxic aero emissions.[159]

Jaguars started repopulating the Argentinian Wetlands after seventy years.[160]

World's first home hydrogen battery launched.[161]

A group of Yemeni women spearheaded the creation of a solar microgrid for a community near a war zone.[162]

Seville, Spain, repurposed robust orange tree harvests to turn into electricity.[163]

The sixteenth Oregonian wolf journeyed to California, showing promising signs of wolf recovery on the West Coast.[164]

Secretary of the Interior Deb Haaland announced the quadrupling of the bald eagle population since 2009 due to conservation efforts.[165]

Nepal's rhino population reached the highest it has been in over twenty years due to habitat regeneration.[166]

Giant pandas are no longer endangered due to Chinese conservation initiatives.[167]

63,000 pounds of trash were removed from the Pacific Ocean through The Ocean Cleanup.[168]

Cuba established a new marine protected area of 281 square miles.[169]

Doctors in Canada can now prescribe patients with trips to National Parks.[170]

Tiger populations are showing an increase after over a century of decline.[171]

In a twenty-two-month study, researchers found that several coral species could survive in ocean temperatures hotter than once previously thought.[172]

Due to extensive reforestation efforts, Scotland's forests are the largest they have been for nine hundred years.[173]

Scientists have observed a 30 percent average reduction in plastic pollution on Australian coasts.[174]

One of the world's largest underwater canyons, located about a hundred miles from New York City, is set to be established as a marine sanctuary.[175]

Turkey discovered nearly seven hundred million mt of rare earth element reserves.[176]

Areas of the Great Barrier Reef saw the highest level of coral in the past four decades, proving the resiliency of the coral ecosystem.[177]

220,000 pounds of plastic were removed from Great Pacific Garbage Patch.[178]

CHAPTER 8

CLIMATE OPTIMISM IN THE COURTROOM

—

One of the most badass things that I have gotten to do thus far in my activism journey is filing a legal complaint against my alma mater—Vanderbilt University—for the administration's failure to align their investments with their mission statement by investing in extractive fossil fuel companies.[179] This experience was one of the most invigorating and motivating campaigns that I have engaged in thus far as a climate activist. It taught me about the importance of fighting from within and outside of the "system."[180]

—

The court of law has been a "petri dish" for change long before issues pertaining to the climate came into play. Several of the laws that many consider as American as apple pie emerged out of litigation once seen as radical.

Let's take it back to a typical high school history classroom in America. For example, in 1806, organizing for higher wages was determined to be a criminal conspiracy under the common law. Thirty-six years later, the groundbreaking Massachusetts Supreme Court decision *Commonwealth v. Hunt* decriminalized unionization and paved the way for American labor movements.[181]

And in a history lesson that many Americans will know well, *Brown v. The Board of Education* used the equal protection clause of the fourteenth amendment to ban the practice of "de facto segregation" within the United States. Before this SCOTUS (Supreme Court of the United States) decision, school segregation was permitted as long as public and private institutions were "separate but equal." *Brown v. Board* exposed that separating students by race within the school system created inherent unequal learning environments.[182]

Despite a delayed start, since the first climate lawsuit was filed in the United States in 1986, a flood of climate litigation has emerged as a powerful and visible way for citizens and organizations to hold corporations and governments accountable for their role in the climate crisis.

As of May 2021, over 1,800 climate change litigation cases have been identified globally.[183] From 2015 to 2021, the number of climate change cases has more than doubled.[184] In case you were wondering, such litigation is described as "cases brought before judicial or investigatory bodies that raise an issue of law regarding climate change as a significant issue."[185] Scholars attribute this rise, in part, to the collective disdain for the lack of action and ambition despite international pledges like the Paris Agreement.[186]

Bringing environmental issues into the courtroom, whether on behalf of youth litigators, marginalized

communities, or nature itself, has become a central way to advance climate action.

It has also been pivotal for citizen mobilization, taking arguments from the streets to the courts.

In a conversation with one of my role models and a badass environmental lawyer, Ted Hamilton emphasized the importance of legal activism, stressing that "we interpret a lot of social problems through court through laws. So if we want this to be a pressing matter of social concern, it also has to have a legal dimension."[187]

Several legal tools and frameworks within climate litigation have been leveraged globally and nationally to protect the planet and its people. Many of the cases using these frameworks are strategic, aiming to establish precedent and evoke a larger cultural or societal shift through the publicity of a legal case. Although many of these novel legal arguments have not yet gained momentum across every inch of the globe, the trend toward consideration of these arguments is certainly promising.

One reason I believe in Climate Legal Activism as a revolutionary, hope-giving strategy is because of its functionality in challenging our conception of justice, undermining the status quo that has for far too long been used to justify the exploitation of the planet and its people.

Where legislative action and market trends fail, defending our planet in courtrooms provides a unique opportunity for radical change.

If you aren't yet as fired up as I am about legal climate activism, I assure you this will help. Here are some examples of successful climate litigation cases from around the world.

SUCCESSFUL CASES

LEGHARI V. FEDERATION OF PAKISTAN—PAKISTAN
In 2015, farmer Asghar Leghari argued that the Pakistani government failed to carry out the National Climate Change Policy of 2012 and the Framework for Implementation of Climate Change Policy. He successfully convinced an appellate court that this governmental failure had detrimental and severe impacts on his right to life and Pakistan's water, food, and energy security.[188]

As a result, the court ordered the government to take action and from September 2015 to January 2017, 66 percent of the priority action items within the Framework for Implementation of Climate Change Policy were implemented.

HAWAII V. HAWAII WILDLIFE FUND—UNITED STATES
In 2020, the SCOTUS found that the county of Maui violated the Clean Water Act because of its wastewater treatment in Underground Injection Control (UIC) wells. The county—like many other US municipalities—used a popular method of injecting effluent (wastewater from factories) into wells even though it is well-known that this chemical "disperses vertically and horizontally, eventually migrating to the ocean." Ninety percent of effluent enters the sea through diffuse flow. The Clean Water Act requires permits for the discharge of pollutants in "navigable waters," and until this case, it was unknown whether underground wells were included within this definition.[189]

This case was hailed as the "Clean Water Case of the Century" by EarthJustice. By including underground wells within the definition of "navigable waters," it became much more difficult for both private and public entities to pollute US water systems.[190]

MILIEUDEFENSIE ET AL. V. ROYAL DUTCH SHELL—NETHERLANDS

In the spring of 2019, a Netherlands-based environmental group (Millieudefensie/Friends of the Earth Netherlands) summoned Royal Dutch Shell (and the entities that belong to the Shell group) to court on behalf of their dereliction of duty of care under Dutch law.

This case was backed by 17,000 co-plaintiffs calling on Shell to reduce its emissions by 45 percent (from 2019 numbers) by 2030 across Scope 1, 2, and 3 greenhouse gas emissions.

> **Scope 1, 2, 3 Greenhouse Gases:** Direct emissions, indirect emissions from purchased energy, and indirect other emissions.

The monumental decision, which was built on the Urgenda case that we will investigate later, paves the way for using the law to make corporations accountable to citizens.[191]

GLOUCESTER RESOURCES LIMITED V. MINISTER FOR PLANNING—AUSTRALIA

In 2017, mining company Gloucester Resources Limited sued the minister for planning for the denial of a new open-cut coal mine in New South Wales.[192] The minister for planning primarily denied the license due to the negative visual and planning impacts of the mine, but in the subsequent appeal, a community group added that the government decision was also justified because of the adverse environmental impacts of the proposed mine. The chief judge on the case ruled in favor of the minister for planning, deciding that the construction of the mine was against public interest for environmental and social reasons, thus setting a precedent for future mine approvals. The judge also rejected the company's argument that if they did not construct the mine, another company in another country would, resulting in the same amount of greenhouse gas emissions. This judge's opinion on the aforementioned argument—called the market substitution defense—is immensely important given the frequent usage of the argument by oil and gas companies around the world. This case marks the first time a coal mine was rejected based on climate change measures.[193]

GBEMRE V. SHELL PETROLEUM DEVELOPMENT COMPANY OF NIGERIA LTD. AND OTHERS—NIGERIA

In this case from 2005, a community representative argued that Shell's practice of gas flaring in the Niger Delta violated the Iwherekan community's rights to human dignity as enshrined under the 1999 Nigerian Constitution.[194] The federal high court honored the plaintiff's arguments, assuring that the constitutional rights to human dignity included the rights to a clean, pollution-free environment.

Even more important than these wins is the trajectory we are observing within legal environmental activism and how it moves the needle on climate action. These legal strategies, even if not successful in winning cases, are successful in establishing precedent—the golden ticket. They are helping shift the cultural norms that indeed frame our laws.

The mere fact that these cases are heard signals the move toward a greater understanding of our place in nature, the urgency of the climate crisis, and the inaction of corporations and governments.

Focusing on encouraging trends in this space rather than the explicit wins allows us to root our optimism less in outcomes and more so in potential. Through extensive research and conversations with experts in the field, I have found that the following legal activism strategies are part of a promising toolkit of change.

THE RIGHTS OF NATURE

Although granting rights to a nonhuman entity might seem radical, it is far more common than you would think. Our country has viewed corporations as persons with rights under the fourteenth amendment since the 1880s.[195] The monumental Supreme Court decision colloquially referred to as Citizens United reinforced this, granting corporations the right to donate to political campaigns without restrictions or regulations, in the name of free speech.[196] Ironically, the decision unleashed the fury of virtually unlimited lobbying on behalf of oil and gas companies, furthering the climate crisis and politicizing climate action. I digress.

In 2008, Ecuador became the first country to give nature constitutionally protected rights under what is more colloquially known as "Rights of Nature," and in Ecuador, "Rights of Pachamama (Mother Earth)." This doctrine confers natural ecosystems with the rights to exist and flourish, independent from interference from industry. As a result, entities protected under Rights of Nature are entitled to legal representation.

Ecuador led the world as the first state to formally recognize the Rights of Nature, largely due to the country's status as a plurinational state[197] (many nationalities coexisting together) that incorporates Indigenous voices and traditional environmental knowledge. The country's highest court recently recognized the prerogative for Indigenous communities to have the final say regarding approval for extractive projects on their land. The uplifting of Indigenous communities and tribes, which view nature wholly different from the dominant, extractive Western lens, is central to the widespread adaptation of Rights of Nature.

The combination of the recognition for Indigenous communities and the establishment of Rights of Nature within the country's constitution has led to landmark decisions.

And it's not just Ecuador. Several countries have adopted Rights of Nature legislation, including Panama, Bolivia, Costa Rica, Bangladesh, Brazil, Colombia, and Uganda.

Resolutions declaring the Rights of Nature have been passed in states and cities across the world. In the United States, Rights of Nature ordinances have been passed in more than thirty local governments.[198] The first community to pass such laws in the states—Tamaqua, Pennsylvania—is a notably conservative area where 70 percent of citizens voted for Trump in the 2016 Presidential Elections.[199] This community has since been followed by Pittsburgh, Pennsylvania; Lafayette, Colorado; Exeter, New Hampshire; Broadview Heights, Ohio; Mexico City, Mexico; Bonito, Brazil.[200]

A country's ecosystems can more specifically be protected under Rights of Nature doctrine on a case-by-case basis.

Several bodies of water have been granted personhood, protecting them against the whims of industry, including the Whanganui River of New Zealand, the Ria Altrato of Colombia, Sukhna lake in India, the Klamath River of California, United States, Lake Eerie of Ohio, United States, those of Orange County, Florida, rivers in Bangladesh, the Mar Menor lagoon of Spain, and Lake Tota in Colombia.

Other ecosystems protected due to recent uptick in Rights of Nature court cases and legislation include the Tu Urewera National Park in New Zealand, and natural entities ranging from the air itself to meadows and grasslands in the Indian state of Madras.[201]

Tribal nations across the world have also included explicit Rights of Nature language into their constitutions, as the framework is born out of Indigenous values and belief systems.

The concept has recently been brought to a US tribal court, with the case of *Manoomin et al. v. Minnesota Department of Natural Resources et al.* In this case, wild rice sacred to the White Earth Band of Ojibwe filed a lawsuit against the Minnesota Department of Natural Resources for their approval of an Enbridge pipeline that threatens to destroy Manoomin's natural habitat. Their case rests on the assertion that Rights of Nature are directly tied to decades-old tribal treaty rights to hunt, fish, and gather on traditional lands.

Despite its current tenuous standing (as of this writing), this case could set a groundbreaking precedent for tribal sovereignty and thus, environmental protection. If the judge rules in favor of Manoomin, it could establish a precedent of extending treaty rights of Indigenous people to extraterritorial land, according to Katie Surma, a journalist and attorney who has extensively covered Rights of Nature legislation for the past few years.

Surma talked with me about the wide breadth of places that have implemented Rights of Nature legislation—from rural US communities to tropical Central American cloud forests—which is promising in its own right, as it speaks to the apolitical desire many communities have to protect their right to a safe environment.

This trend, I believe, might be most impactful in the cultural shift it necessarily brings about. By giving rights to natural ecosystems, this legal argument necessarily expands the definition of rights and calls out the injustice of prioritizing human over nonhuman rights.

REASON FOR HOPE: THE DUTY OF CARE

One thing all politicians can agree on is the duty of the government to protect its citizenry. Just how one defines protection and ensures the citizenry are protected might be disputed, but the sentiment and political will to do so persists across the board. Although governments have promised protection through a slew of international treaties, treaties lack enforcement. Due to the nonbinding nature of international climate agreements like the Paris Climate Accords, people have turned to legal measures to hold governments accountable to this promise of protection.

The duty of care argument was applied to protect children from the consequences of climate change for the first time in the landmark *Urgenda Foundation v. Government of the Netherlands (Ministry of Infrastructure and the Environment)*. In this case, The Hague ruled in favor of the argument that the current climate policies of the government were insufficient in fulfilling the duty of care to Dutch citizens.

This marked the first time citizens successfully argued for the governmental duty to prevent climate change.

Based on the argument by the Urgenda Foundation, the District Court of The Hague ruled that the government must slash greenhouse gas emissions by at least 25 percent by the end of 2020.²⁰² The government filed several appeals in attempts to reverse this ruling, but in doing so was only able to delay their prescribed timeline.

The Dutch government argued that this decision would symbolize an overreach of the judiciary, an argument frequented by municipalities all over the States, but the court smacked this down, stating "the task of providing legal protection from government authorities, such as the State, preeminently belong to the domain of a judge."²⁰³

The Urgenda Foundation supported their argument by invoking rights-based arguments (which we will explore soon), although The Hague District Court did not find any human rights violations in the case.

The precedent established by this case has empowered youth and citizen groups around the world to stand up for the governmental duty to protect (read: care for) its citizens and future generations, emphasizing the injustice of relegating youth to a sordid future before they even have the right to vote for the future they want.

After 58,000 Belgian citizen co-plaintiffs came together to assert this same argument in VZW Klimaatzaak v. Kingdom of Belgium & Others, the court ruled in partial favor of the plaintiffs, finding that the federal state and other regions invoked "[were] jointly and individually in breach of their duty of care for failing to enact good climate governance."²⁰⁴

The Urgenda decision also inspired another Dutch foundation to use the argument, but this time to hold corporations accountable. If you remember *Milieudefensie et al. v. Royal Dutch Shell* (mentioned earlier), this ruling set requirements for Shell's future emission cuts. As a result, the court extended the duty of care argument to corporations, marking a huge leap in the court's discretion of how corporations should or should not protect citizens of a country.

The duty of care doctrine, as it pertains to the environment, gives me hope because it crystallizes the idea that the government and even, in some cases, corporations have a duty to preemptively protect citizens from natural disasters and planetary destruction.

SPOTLIGHTING ECOCIDE

The image of orange smoke plumes emanating from fighter jets during the Vietnam War, dispelling chemical-laden Agent Orange on foliage, is what one might think of when they think of **ecological genocide, or "ecocide."** Though the term was first used to refer to this inhumane practice, the concept now includes a wide range of environmental tragedies, both seen and unseen.

Forty-nine years after the dumping of Agent Orange during the Vietnam War, the concept of ecocide was presented to the International Criminal Court by the island nations of Vanuatu and the Maldives.[205]

The big kahuna of all criminal courts, the International Criminal Court (ICC), investigates four types of crime identified as the most heinous of all: genocide, war crimes, crimes against humanity, and the crime of aggression. Because of its international scope, any crimes within those categories waged against a country that has signed onto the Rome Statute (the treaty establishing the ICC) are subject to international punishment. You don't want to get on the wrong side of the ICC.

The ecocide movement, championed by figures like the Pope, calls for the criminalization of ecological genocide in the highest court of law. Although environmental destruction was once a part of the earliest drafts of the Rome Statute, the enshrining document of the ICC, it was soon removed due to opposition from the United States, United Kingdom, and the Netherlands.

Since then it has been ideated upon by concerned citizens and activists all over the world. Recently, a team of international legal experts, as part of the ecocide drafting panel, came up with a definition of ecocide to be used in the courtrooms. They describe ecocide as "unlawful or wanton acts committed with knowledge that there is a substantial likelihood of severe and either widespread or long-term damage to the environment being caused by those acts."[206]

The term has been invoked several times by Indigenous groups in Brazil, accusing the Brazilian administration under Bolsonaro of waging ecocide against the Amazon.[207]

And though it has not yet been amended to the Rome Statute, the ICC took a significant step in a 2016 internal policy paper by including the destruction of the environment under the umbrella of wartime offenses.[208] Additionally, national ecocide laws have emerged in several countries, including Russia, Kazakhstan, Kyrgyz Republic, Tajikistan, Georgia, Belarus, Ukraine, Moldova, and Armenia.[209]

Significantly, if ecocide became the fifth internationally recognized crime, each ICC member state—all 123 of them—would be required to enable national ecocide laws. If they failed to enforce these laws, the ICC would have the prerogative to hold them accountable. It is important to note that four of the world's biggest polluters: China, the United States, India, and Russia, are not parties to the Rome Statute. Nonetheless, if these countries have corporations operating within a Rome Statute member state, they could still be held accountable.

Despite it being a prosecutorial strategy, the goal of the movement is to deter large-scale environmental harm by governments or corporations.

Jojo Mehta, the Executive Director of Stop Ecocide International, told me about the remarkable momentum the movement has received in the last few years. Mounting public discourse on ecocide and its implications has brought the number of countries expressing support (either through legislative proposals or statements) for the criminalization of ecocide from zero in 2019 to at least twenty-four, at the time of writing.

This strategy is uniquely encouraging because of its international scope. While other legal strategies are bound by a judge's interpretation and often subject to reversal, an international criminalization of ecocide would clarify the culpability of pervasive polluters.

Ecocide can transcend international borders, addressing the international scope of the climate crisis. Criminalizing ecocide can address the environmental injustices that have been waged against Global South countries perhaps better than any other legal strategy.

As Mehta said, "Most, but not all but most of ecocide or type activities take place in the developing countries and in the Global South, whereas the decisions that lead to them almost invariably are taken in the wealthy North. And the beauty of the ICC route is that it aims at individuals in positions of superior responsibility."

A REASON FOR HOPE: RIGHTS-BASED CASES

Much of the basis of the rights-based cases revolve around protecting human rights where they have been violated, according to a nation's constitution. In the US, rights-based cases typically deal with the protections granted to citizens under the fourteenth amendment. The right to marry was extended to same-sex couples (*Obergefell v. Hodges*), as was the right to receive a quality education, regardless of race (*Brown v. Board of Education*).

In an increasingly popular legal argument, climate change activists are bringing the rights-based argument to climate change litigation. This legal strategy is undergirded by the objective truth that climate change is already violating human rights all over the world by forcing people to migrate from their homes, depriving children of the ability to get an education, and making it difficult for low-income, marginalized communities to obtain necessities like clean drinking water or breathable air.

Leghari v. Federation of Pakistan, which we touched on earlier, was a landmark case for the rights-based climate litigation strategy.[210]

Soon after the Leghari case, in 2016, twenty-one young plaintiffs filed suit against the US Federal Government in *Juliana v. United States*.[211] The plaintiffs argued that the government's endorsement of the fossil fuel industry and subsidizing of oil and gas companies directly fueled dangerous carbon dioxide concentrations and violated their rights to a stable climate system.[212]

The judge denied the defendant's motion to dismiss the suit, and the case proceeded to trial. Despite the reality that the most recent ruling by a circuit court nullified standing due to a lack of redressability (the court found the relief sought by plaintiffs to be outside of their power),[213] this case was significant in that it pushed forward the rights-based argument despite consistent attacks from industry and the executive branch, and was declared justiciable (within the court's judicial authority) by the highest court in the US.[214] As of this writing, the plaintiffs are awaiting a ruling from the US District Court Judge Ann Aiken on a motion to amend and hope to return to trial soon.

Legal scholars project that the most successful rights-based cases will be regarding "domestic constitutional rights litigation challenging government mitigation or adaptation failures; secondly, cases in European jurisdictions following the Urgenda model of using rights to question the adequacy of emissions reduction targets; and, thirdly, possibilities for human rights climate change claims under regional human rights treaties operating outside Europe."[215]

The rights-based approach addresses the consequences of climate change on human life and is significant in updating how a country thinks about its citizens' rights and what they include. Despite the many legal loopholes that arise when using this strategy, it gives us hope for a world where environmental rights are enshrined in our sacred documents.

DAMAGES AND FRAUD CASES

The oil and gas industry has its fair share of friends, which notably (and unfortunately for many democracies) include the politicians elected to serve us. The brigade of oil and gas friends is often referred to as the Climate Change Countermovement (CCCM).[216] Coalitions within this countermovement "pool resources from a large number of corporations and execute sophisticated political and cultural campaigns designed to oppose efforts to address climate change."[217] Many of these aforementioned campaigns have been riddled with fraud and deceit to pursue profits at the cost of the planet.

The American Petroleum Institute, perhaps the most nefarious and largest of the CCCM members, was aware of the direct relationship between the burning of coal, oil, and natural gas and global warming in 1965.[218]

Despite knowing the responsibility held by the industry to control global warming, companies engaged in intentional deception and fraud. The tactics employed by these companies to both obscure the ramifications of their work and to deceive customers (launching massive PR campaigns and commissioning studies to cast public doubt about the threat of climate change, etc.) rival that of the tobacco industry.

A content analysis of ExxonMobil's communications on climate change from 1977 to 2014, including internal documentation, advertorials (paid advertisements in the form of editorials), and publications, showed that while 80 percent of the internal company documents acknowledge climate change as real and human-caused, only 12 percent of the thirty-six advertorials they published within this period did the same.[219]

This is perhaps best exemplified in an advertorial ExxonMobil Corp took out in *The New York Times* in 2000. Titled "Unsettled Science," the writer (paid by the company) asserted that "scientists remain unable to confirm either contention [that humans are causing global warming and that dangerous impacts are already under way]."[220]

To be clear, ExxonMobil is far from the only oil giant culpable of misleading the public and driving climate inaction.

The systematic dismissal of scientists' concerns and shrouding of their role in the climate crisis is not a relic of the past. Although these companies now by and large acknowledge climate change (not doing so now would be embarrassing), they are engaged in similar shenanigans.

Despite the glittery promises of a carbon-free future they are supposedly building, the industry's new modus operandi is greenwashing at its finest. Research shows that despite a significant increase in the mentioning of "low-carbon transition[s]" and hefty net-zero goals, oil and gas companies fall short when it comes to concrete actions that would substantiate these pledges and goals.[221] In fact, of the 2021 capital expenditures of seven top oil and gas companies, less than 15 percent of CapEx (capital expenditure) was invested in low-carbon energy.[222]

Because of the increasing publicity highlighting these tactics and the connection between this industry and the tobacco and opioid industries, a mounting suit of litigation is being filed against members of the CCCM for damages/liability and fraud.

Lawsuits against oil and gas companies typically fall under these two categories. **Damages cases**, typically waged by cities or states, address the impact that oil and gas companies have had on the climate crisis and, as a result, the mounting costs cities face in terms of climate adaptation. **Fraud cases** are typically put forth by governments or shareholders who believe that they have been misled by a company's lack of disclosure of the risks and realities of climate change.

Since 2018, damages and fraud cases have popped up in states and cities across the US, from Minnesota to California to South Carolina to Maryland to Hawaii, as the costs (both financial and human) of climate change and industry fault become increasingly clear.

In City of Annapolis v. BP p.l.c., Annapolis, which has the greatest increase in average nuisance flooding events of any city in America,[223] alleged the deception tactics and greenwashing of more than twenty oil and gas companies have resulted in mounting costs of climate change mitigation. In the words of Annapolis mayor Gavin Buckley, this lawsuit "shifts the costs back to where they belong, on whose knowledge, deception and pursuit of profits brought these dangers to our shores."[224]

The California attorney general launched an investigation into the fossil fuel industry's role in causing the plastic pollution crisis in April 2022, paving the way for a potential fraud case that addresses a different impact of the fossil fuel industry.[225]

A couple of these cases have made noticeable headway in the path to go to trial, despite the legwork the CCCM has put in to stop them.

On the other side of the world, another fraud case is brewing in an Australian court, led by shareholders suing the Commonwealth Bank of Australia for its failure to include climate risk in its risk management framework.[226]

Individuals too can play an instrumental part in fraud and liability litigation. In a securities fraud case—*Ramirez v. Texas*—argued by an Exxon investor himself, the Exxon investor accused the company of artificially inflating the stock price, due to the omission of climate risk disclosure.[227]

To be transparent, we are fighting an uphill battle by taking these all-too-powerful companies to court with their expert and expertly paid legal teams (the top hundred law firms in the US facilitated $1.36 trillion in fossil fuel transactions in the last five years).[228]

Although these cases are relatively nascent, I am hopeful about this trend of litigation, which represents mounting pressure on these powerful companies.

These cases are expensive for oil companies due to litigation costs and the negative press attention that results in increased shareholder activism and citizen backlash. These cases also have a greater chance of succeeding in the court system due to the improvements in attribution science, which help to link specific sources of pollution and emissions to specific climate calamities.

CLIMATE NECESSITY DEFENSE

Environmental activists have long been villainized by corporations and governments as aggressive and obstinate inhibitors of "development." Individuals and groups who have done nothing more than trespassing and turning off pipeline valves reside on the Department of Homeland Security "extremist" list, alongside the likes of mass murderer and white supremacist Dylann Roof.[229]

The climate necessity defense is increasingly important in the wake of this criminalization. The legal strategy is a defendant-focused approach to absolving environmental defenders and activists of crimes. Like the duty of care doctrine, the necessity defense doctrine is nothing new. This common law doctrine allows defendants to argue that any crime they have committed was to prevent a bigger crime from happening. But the usage of this doctrine in environmental contexts certainly is novel, a response to both the increasing criminalization of environmental defenders and the increased urgency of the climate crisis.

This defense hinges on the argument that civil disobedience and nonviolent direct action are necessary when the government fails to protect citizens from the harms of climate change and all other legal alternatives are exhausted. In the US, this argument has primarily been used in reference to interference with energy infrastructure like pipelines. As succinctly put by the members of the Climate Defense Project, the climate necessity defense asserts that "oil pipelines are more dangerous than those who protest them."[230]

In addition to fighting the day-to-day consequences of climate change that manifest in more frequent natural disasters and higher temperatures, environmental activists are standing up against the slow violence waged against them by governments and corporations as a function of capitalism and extractivism. The collisions of activists and governments/corporations often result in environmental conflicts.

The Environmental Justice Atlas catalogs 2,743 environmental conflicts around the world. Most of these conflicts are nonviolent, but that does not stop the presence of the police state in these conflicts. Of the cataloged cases, 20 percent resulted in the criminalization of defenders, 18 percent in physical violence, and 13 percent in assassinations. These numbers are higher when Indigenous people are involved (41 percent of cases).[231]

In recent years, these activists have encountered an increasingly hostile response from corporations and members of government. In the US, state governments have implemented Draconian punishments for constitutional protesting around "critical infrastructure"—or fossil fuel infrastructure. Since 2017, thirty-eight anti-protest laws have been enacted across twenty states.[232] One such anti-protest law in Alabama prohibits nonprofits from organizing or even providing support for environmental protests near critical infrastructure. Punishment for violation falls under a Class C felony, which is punishable by up to ten years in prison.[233]

Laws like this one from across the country come so close to infringing on constitutional rights that they have been flagged by officials at the United Nations as having a "detrimental impact on the rights to freedom of peaceful assembly and freedom of expression in the country."[234]

An example of this legal strategy plays out in the efforts of the rag-tag group of climate activists called the "Valve Turners." Known for organizing a cross-country effort to shut off pipelines carrying tar sand oil from Canada to the US, this group of five activists was some of the first to bring the climate necessity defense into recognition.

Significantly, this case was the first instance in which a jury heard the climate necessity defense, despite a decisive set of attempts by prosecutors to block the presentation of the argument.

Similarly, in *Massachusetts v. Gore*, thirteen people laid down across the West Roxbury Lateral Pipeline to protest its construction. After employing the necessity defense, the defendants were acquitted on charges of trespassing and disorderly conduct.[235]

In *State v. Delahale*, defendants who removed a portrait of French President Emanuel Macron as an act of protest against his inaction on climate successfully argued the necessity defense and were acquitted of robbery charges.[236] Even though this crime was more symbolic (the physical removal of the French president's portrait is certainly not directly preventing greenhouse gases from entering the atmosphere), its standing in court attests to the broadening acceptance of activists' efforts.

South of the equator, the argument—despite going by other names—is also emergent, especially as it pertains to Indigenous land defense.

The largest trial in Peruvian history followed the 2005 Baguazo massacre, where a fifty-day-long road blockade by Indigenous Amazonian communities in efforts to prevent a forestry law that would privatize the most biologically rich part of the Amazon basin resulted in thirty-eight deaths (between police members and Indigenous community members).[237]

Passage of this forestry law, despite Indigenous resistance, would violate the International Labour Organizations Convention 169, which was ratified by the state in 1994.[238]

Seven years later, the case involved fifty-seven Indigenous leaders facing charges for the death of policemen during the protests.[239] In a monumental decision, Peru's courts ruled to clear fifty-two Indigenous leaders of all homicide charges and charges related to blocking the highway, referencing that these community members were acting out of necessity to defend the environment.[240]

Like many of the other strategies previously mentioned, the promise of this strategy doesn't lie in its potential to succeed in court. If defendants are given the chance to present this defense to the jury, protesters can explain the reasoning, motivation, and need for nonviolent direct action in the wake of dreadful climate inaction by legislative bodies or corporations.[241]

As time shows, those with the money and power are not taking the action they need to stop the raging boulder of the climate crisis heading toward us at full force. The traction of the climate necessity defense is encouraging because it asserts certain validations on behalf of the court; that governments and corporations are failing to take adequate and timely action, that property law does not necessarily protect people, and that environmental defenders are justified.

RECAP

All these legal strategies are undergirded by a larger cultural shift that places a value on nature that, in the dominant Western perspective, has only been granted to some people. Climate change litigation is happening all over the world, even though those in the Global South get remarkably less attention.[242]

The domino effect of these legal arguments, many of which call on similar precedents in argumentation, should be emphasized. Despite many cases not being "successful" in the traditional sense, the fact that many novel legal strategies are gaining traction across these cases is certainly significant.

CHAPTER 9

STARTING A MOVEMENT

—

> "SOCIAL MOVEMENTS ARE AT ONCE THE
> SYMPTOMS AND THE INSTRUMENTS
> OF PROGRESS."
>
> —WALTER LIPPMANN

My first memorable involvement with a movement was not entirely climate-related, but just enough so that it felt comically appropriate to include in this chapter. I took Advanced Placement Environmental Science my senior year of high school, the first year it was being offered, and I distinctly remember an uprising (not using that lightly) during the first week of class.

In protest against our teacher's pop quiz and in true high-schooler fashion, two-thirds of the class marched to the guidance counselor's office after last period, whining over each other about how unfair this quiz was.

Truthfully, I wasn't too upset about the quiz. Not to toot my own horn, but I did well on it—a good omen considering this subject became my undergraduate major and eventual career. But despite not being passionate about negotiating the quiz, I rallied behind my peers and stood in the back of the uncomfortably crowded counselor's office, invigorated by the energy in the room. I have since participated in many more movements, significantly more meaningful, but still carrying an energy reminiscent of my first.

When thinking about movements, people typically think about the tip of the iceberg (climate pun indeed intended), the swarms of people marching in the streets. But movements are far more than this. They include uniting different groups around a common cause, lobbying politicians, educating people, proposing policy, putting pressure on powerful entities, and shifting the larger discourse on an issue.

Every social advancement has at least been in part due to the efforts of social movements. Movement building is a broad term—and can include everything from community-based organizing to regime change. But most movements start at the grassroots level, involve some sort of theory of change, and operate to advance a certain agenda or perceived set of rights. There is a long history of grassroots activism resulting in real, tangible change—whether in the government or society at large.

One such example is the Civil Rights movement in the US. Within a ten-week period in the spring of 1963, around 758 public demonstrations for desegregation were held throughout the country.[243] The magnitude of protests effectively pushed President John F. Kennedy to announce proposed strategies far more progressive than those anticipated at the start of the year.[244]

That critical mass of people to make a movement successful seems intimidating. But research of movements aimed at toppling repressive governments shows that the movements practicing nonviolent civil resistance only needed the active support of 3.5 percent of the population to succeed in advancing their movement goals.[245] So imagine what that number must be for a little less intimidating task, like getting a government to transition to clean energy at a faster rate.

Despite the limited amount of research regarding the impact of activism in the environmental space, there is evidence that institutional protest activities significantly raise the amount of Congressional hearings on the environment[246] and that emissions in states decline in relation to the volume of pro-environmental protests.[247]

In this section, let's look at environmental movements that largely emerged from **grassroots organizing**, a term largely used to describe activism that starts at the community level, which I believe is the lifeblood of democratic change.

Whenever I have a conversation with someone above fifty about environmental activism, they always reference the first Earth Day in 1970. This day was the largest mass mobilization in American history, with some estimates counting twenty million participants attending 12,000 events around the country.[248] It was born of an era ushered in by counterculture—you know, the hippie, peace-loving, anti-war, Woodstock-going type that I think I would have thrived alongside. It rode on the coattails of the Civil Rights Movement, which was (and remains) far from over. Though a correlation is not certain, it is likely no coincidence that the Environmental Protection Agency was established by Executive Order the following July.

The following decades have been marked by international summits, growing outrage, and youth mobilization.

Although the first yearly United Nations Climate Change Conference was in 1995, researchers regard the 2009 UNFCCC COP15 summit in Copenhagen as a turning point for climate movement mobilization.[249] Organizational presence was unprecedented, with two-thirds of the registrants being NGO (non-governmental organizations) observers.[250] People signed up for the conference in record enrollment and calls for climate justice were louder than they have ever been at previous summits.

This conference was marked by over 100,000 protestors, a new rallying call for climate justice, and the usage of outsider tactics like protests.[251] Lackluster action followed the conference and a new wave of disenfranchisement ensued.

Fridays for Future (FFF) marked a turn in the global climate movement. First characterized by Greta Thunberg skipping school to take to the streets on—you guessed it—a Friday, the organization soon became global in scope, mobilizing a high of 7.6 million people around the world.[252]

Soon after, Extinction Rebellion (XR) was born with the issuance of a Declaration of Rebellion against the UK government, accusing the government of shattering democracy and breaking its law in favor of short-term profits and environmental destruction.

Both these organizations differed from the previous dominant narrative of environmental movements by "refocusing on the state."[253] Messaging diverted from that of previous movements, which focused more on individual or lifestyle changes, and turned increasingly toward advocating for governmental change, imploring governments to listen to scientists, keep up with their Paris Agreement commitments, and protect their citizens.

The deflection of blame away from individuals for failing to act on climate and toward corporations, governments, and the wealthiest individuals on the planet, has democratized the environmental movement, bringing more people into its fold.

More recently, with the finding that just a hundred companies are responsible for 71 percent of all greenhouse gas emissions,[254] high-emitting corporations have also become the target of the movement.

Movement building, especially environmental movement building, is encouraging because of the diversity of people involved with it. The diffuse nature of the environmental movement, which has roots in every sector from agriculture to apparel manufacturing to banking, calls for a vast set of opportunities to make a change.

It is difficult to pinpoint what constitutes a "win" within movement building, as victories can range from policy proposals to a change in public opinion to the enactment of pro-environmental legislation. For this book, we will consider wins in the space as movements that have effectively mobilized thousands of people through innovative responses to the crisis at hand.

A REASON FOR HOPE: LANDBACK

Indigenous communities have been advocating for the transfer of governance and power back to Native communities ever since their land was violently stolen out from under them by colonizers hundreds of years ago. But social media coverage from the Dakota Access Pipeline demonstrations popularized the LandBack movement, calling in Indigenous allies and environmentalists from around the world to stand in solidarity with Indigenous people and water/land. The LandBack Manifesto, drafted by the NDN Collective, was published on Indigenous Peoples' Day 2020.[255] The manifesto calls for the reclamation of everything stolen from the original people.

In the last few years, members of civil society have increasingly been attuned not only to the justice component essential to LandBack, but the environmental benefits that the world reaps when we return Indigenous land to its original stewards.

Since 2004, tens of thousands of acres of ancestral land have either been bought back or given back to Native tribes in the United States, including those of the Wiyot Tribe, Mashpee Wampanoag, Wyandotte Nation of Oklahoma, Esselen Tribe, Leech Lake Band of Ojibwe, Passamaquoddy Tribe, and the Salish and Kootenai Tribes.

Falling short of full restitution, other measures toward LandBack have been taken, such as the 2020 Supreme Court decision *McGirt v. Oklahoma*, in which the United States Supreme Court upheld that around three million acres of eastern Oklahoma is the unceded land of the Muscogee Creek Nation.

Nadya Tannous from NDN Collective describes the movement in the last few years as being marked by pathways to LandBack. She distinguishes the different pathways to LandBack—land trusts, co-ops, legislations, lawsuits, land taxes, reclamations—from the concept of land restitution itself, but emphasizes the immense wins by measure of these various pathways in her time working for the cause.[256]

There is a clear moral imperative to return land to the original stewards of it after years of forceful removal, government-sanctioned murder,[257] treaty-breaking,[258] and erasure of Indigenous communities. But there is also a clear ecological imperative to do so to protect our planet.

Indigenous-managed lands have higher levels of biodiversity than protected lands in several countries.[259] Despite only making up 5 percent of the world's population, Indigenous people steward approximately 85 percent of the world's biodiversity.[260]

In addition, Indigenous-led resistance to fossil fuel projects in the last decade is estimated to have stopped or delayed greenhouse gas pollution equivalent to at least one-quarter of annual US and Canadian emissions.[261]

Efforts on behalf of the federal government, states, and municipalities to give Native communities land back will never wipe away the bloodied history of Indigenous relations in the United States. It can, perhaps, imbue us with a sense of hope for the ancestors and descendants of the original stewards of this land we call home to get justice—and in doing so, receive help in their centuries-long fight to save the planet.

CLIMATE JUSTICE + INTERSECTIONAL ENVIRONMENTALISM

Just as Black and Brown families are more likely to be denied equal access to housing, so too are they more likely to live near a landfill or a polluting industrial plant. With the connection that housing, healthcare, and education has to the environment one lives in, it is no surprise these communities face unrivaled environmental health hazards.

Dr. Rob Bullard (commonly referred to as the "father of environmental justice") was one of the first people to study environmental racism in the US. In my hometown of Houston, Texas, Dr. Bullard published a study in 1983, showing that despite the population being only 25 percent African American, 80 percent of city-owned garbage incinerators and 75 percent of privately owned landfills were in Black neighborhoods.[262]

In 1991, Dr. Bullard was part of the seminal conference called the First National People of Color Environmental Leadership Summit, which brought together BIPOC environmental leaders from all over the country. Conference members redefined what the nebulous term "environment" encapsulates, migrating away from its connotation of an untouched, pristine wilderness scene which was often perpetuated by white-led organizations and shifting toward an understanding of the environment as the place(s) that people "lived, worked, studied, played, and prayed."[263] A key takeaway from the summit, which is now a seminal reference in environmental justice circles, is the **Seventeen Principles of Environmental Justice**. These principles helped form the climate justice movement as it is understood today.

The environmental justice movement seeks to address climate change in a way that addresses these inequalities but also roots out the deeper evils that propel the climate crisis, like unchecked capitalism, racism, and the patriarchy. It advocates for addressing the systemic intersections between social justice and environmental issues and the increased representation of marginalized communities within the environmental movement.

But just as racism permeates through the layers of every existing policy and institution globally, resulting in environmental injustice, so too has it through the roots of the nature of environmentalism.

This discussion of climate justice has, for far too long, been sequestered by mainstream environmentalism—a space that was dominated by voices who advocate for the planet but not its people. White men dominate boards of environmental organizations, lead discussions, and get to move the needle on issues. Because of this lack of representation, combined with systemic inequalities that still have not been addressed, the climate justice movement has not been accepted into mainstream environmentalism until a few decades ago.

The Environmental Protection Agency first recognized these environmental disparities in 1990.[264] In the last decade, we have seen climate justice rise to the forefront of conversations and planning, due to the tireless work of organizers and the heartbreaking stories of Cancer Valley and Flint, Michigan, in the United States, two majority African American places where the presence of polluting industries and poor infrastructure, respectively, have caused severe environmental health crises for the communities.

During the 2020 US presidential debates, the moderator asked candidates about environmental racism[265]—marking a transition of this important subject not just into mainstream environmentalism but also into mainstream politics.

President Biden has since formed the first White House Environmental Justice Advisory Council. In response to the council's recommendations, the administration launched the Justice40 initiative, which has committed to earmarking 40 percent of federal investments that fall under the categories of climate change, clean energy/ energy efficiency, clean transit, affordable and sustainable housing, training and workforce development, reduction and remediation of legacy pollution, and the development of clean water and wastewater infrastructure[266] to marginalized groups—those impacted by environmental racism.

This progress is notable, although we have further to go in rectifying the generational injuries inflicted upon Black, Brown, and Indigenous people.

Twenty-nine years after the First National People of Color Environmental Leadership Summit, shortly after the tragic murder of George Floyd in 2020, the American environmental activist Leah Thomas posted a multi-image Instagram carousel with a graphic that read "environmentalists for Black Lives Matter" and a pledge for "advocates for justice for people+planet," whom she called Intersectional Environmentalists.[267] In social media terms, the post "blew up," reaching millions of people. In a time when activism was taking on a whole new digital sphere, and people were looking for ways to get more involved in the fight against rampant injustice, Thomas gave those in the environmental movement an access point to get involved with its inextricable connections to social justice.

The Intersectional Environmentalist (IE) pledge, inspired by the work of American Civil Rights advocate Kimberlé Crenshaw in developing intersectional theory,[268] calls on individuals to stand in solidarity with marginalized groups, use privilege to advocate for oppressed voices, seek out the intersections between social and environmental justice, and "proactively do the work to learn about the environmental and social injustices these communities face without minimizing their voices."[269]

As of this writing, over one million people have shared it.

In that time, the organization has grown from a pledge people sign to a framework environmentalists can adapt to ensure that they are advocating for a planet that is safe for all people. In the words of Sabs Katz, one of IE's cofounders, the growing movement "pulls [together] a lot of people working within racial justice, gender justice, and reproductive justice...and encourages us to understand all of those other injustices and how we can combine our movements to build a future for everyone that is more equitable and just for everyone."[270]

Having firsthand seen the growth of the Intersectional Environmentalism movement, I can attest to its power in calling in people who might not have otherwise been drawn into the realm of environmentalism. For those already entrenched in environmental work, it has pushed us to look for the connections between environmental degradation and other forms of oppression, better situating us to fight against all extractive systems hurting the planet and its people.

I am encouraged by how the Intersectional Environmentalism movement expands the definition of environmentalism—which, for far too long, has been a place where existing systems of power, like the patriarchy and white supremacy, have thrived—to include and amplify marginalized people.

STANDING UP TO FOSSIL FUELS

Eighty-nine percent of global carbon dioxide trapped in our atmosphere today comes from three products—coal, oil, and gas—all of which fall under the category of fossil fuels.[271] With the astonishing footprint of these three resources, you would think that a majority of attention within the environmentalism space has always been directed toward the companies responsible for endless cycles of extraction and burning. You would be wrong.

In the '70s, large-scale activism looked different than it does today. Conversations around environmental action were either led by think tanks and government relation groups, which aimed their efforts at policy change, or individuals advocating for recycling.

The increased antagonism toward oil and gas companies within the environmentalist movement necessarily spurred an increase in grassroots organizing and citizen involvement. No longer was the movement limited to those who were familiar with bureaucratic jargon and processes.

Not only was this change within the movement marked by increased grassroots mobilization, but also by a more targeted and pointed approach to the crisis at hand. Whereas the climate crisis can often seem imperceptible, oil spills, smokestacks, and rigs are intensely visible and visceral. This physical culpability undoubtedly helped accelerate the movement's traction.

Founder of the Fossil Fuel Non-Proliferation Treaty and long-time climate advocate, Tzeporah Berman, attributes the success of the anti-fossil fuels movement in part to its ability to pull people from outside the climate movement into the space.

Berman's involvement with the anti-fossil fuel aspect of the movement started when she learned about the tar sands projects in Alberta affecting Indigenous communities like the Athabasca Chipewyan First Nations.

Her organizing efforts quickly converged with other passionate activists to target behemoth oil and gas projects that threatened to contaminate Indigenous land and water sources.

Emboldened by the realization that each new pipeline would not be the last, and that the larger enemy was the fossil fuel system, Berman and a coalition of environmental defenders decided to target the supplier infrastructure at large.

With the imminent construction of the Keystone XL Pipeline, these activists endeavored to find a way to stop it—and every other destructive tar sands pipeline going forward.

Despite the initial hostility that many in the environmental movement directed toward this supply-side activism, many people from a variety of backgrounds came around to the anti-fossil fuels movement with the realization that challenging the system responsible for creating our reliance on dirty fuels, rather than the consumer themselves, might be the most effective way to fight against the climate crisis.

This momentum has only been accelerated by the wins activists have witnessed in the closing of Energy East Northern Gateway and Keystone XL pipeline in recent years.

The anti-fossil fuel movement is wrapped up in the creation of the Fossil Fuel Non-Proliferation Treaty, a treaty that calls for an end to all new fossil fuel exploration and production, the phase-out of existing production, and the just transition toward a renewable-centered economy.

To date, it has been endorsed by over fifty cities and subnational governments, from London, United Kingdom, Itahari, Nepal, to Los Angeles, USA; 1,250 organizations, from Amnesty International to Islamic Relief Worldwide; and over 150,000 prominent individuals.[272]

Through the endorsement of this treaty, cities, organizations, and individuals around the world, including London, Hawaii, Montreal, and Geneva, express a commitment to halting all new oil and gas lease projects.

The symbolism of a fossil fuel non-proliferation treaty should not be understated. Likening the destructive forces of the fossil fuel industry to that of a nuclear arsenal is powerful, to say the least. Acknowledging fossil fuel infrastructure as a structural issue that must be phased out aligns with the findings of the IPCC and signals the imperative to do so to limit global warming to 1.5 degrees Celsius above preindustrial times.

DEPOLITICIZING CLIMATE CHANGE IN AMERICA

If you had to guess, which American president started the Environmental Protection Agency? I'll give you a moment to take a guess.

President Richard Nixon. I know—not what you thought. What future president in 2009 signed an open letter to President Obama and Congress in *The New York Times*, urging them to make stronger climate commitments in the upcoming United Nations Conference? Donald Trump. Go ahead and reread; that wasn't a typo.

In the time since, partisanship has sunk its teeth into the climate issue, mostly driven by corporate interests (fossil fuel money), solutions aversion, and media bias. In the last few decades, the once bipartisan issue of healing the planet and protecting its people has become extremely politicized, isolating much of an entire political party from joining in on the fight for our planet and its people.

But the fact that it once was an apolitical cause means it can be again, and the growing body of research on effective climate communications to bridge political divides paired with the trend of conservative youths stepping up to urge their party to take action on climate change is encouraging for the healing of our fractured political system that is particularly ineffective on this issue.

To best address a problem, we must accurately define the scope, threat, and depth of the problem, identifying its root causes. Climate science is just that—science. So the distrust of science that climate change deniers pedal is quite curious. But behind a layer of skepticism that perhaps inhibits denialists from reading and believing in science, lies a likely fear of the solutions proposed to ameliorate the crisis.

Solutions aversion, the lack of acceptance of a problem due to the solutions it implies, often undergirds the partisan divide on climate change, especially due to the policy-oriented call to action that climate change necessarily brings. Essentially, the fear of one-sided policy solutions to the climate crisis has led politicians to sow doubt around the fact that the climate crisis exists.[273]

This research is instrumental in learning to move past partisanship and push for meaningful and impactful policy change. Climate solutions that are not completely one-sided and incorporate values and strategies from both sides of the aisle should be increasingly pursued. While we cannot afford to skimp on climate policy, we especially cannot afford to neglect it altogether due to a lack of agreement. The Bipartisan Senate Climate Solutions Caucus, made up of fourteen senators from all over the geographical and political map, including Senator Lindsey Graham, is a great example of this.

One powerful policy that has support from voices on both sides is a federal carbon tax—for the Democrats, it penalizes polluters without letting them pay their way out of accountability, and for the Republicans, it is a market-based solution that can pass benefits onto consumers. At this time of writing, there have been at least nine acts introduced by members of Congress on both the right and the left that incorporate carbon taxes.[274] There remains a host of issues surrounding public opinion on carbon taxes, but this potential site for bipartisan change should be explored more.[275]

The growing body of research on climate communication also helps us understand how to better talk to people across the political aisle about climate change. For example, a fascinating study shows that framing environmental issues in a past-focused manner resonates with conservatives, whereas framing environmental issues in a future-focused manner resonates with liberals.[276]

While partisan division on climate change runs deep, young people within the conservative party are standing up to older members of the party—like Mitch McConnell—who are struggling to recognize the severity and imminence of the crisis. With the grit and ferocity innate in young climate activists, these conservatives are moving the needle on this important issue within their party.

Benji Backer, the founder of the American Conservation Coalition, is no stranger to the partisan tensions on climate change. His organization, which advocates for primarily market-based climate solutions, created the American Climate Contract, a framework based on energy innovation, global engagement, natural solutions, and infrastructure. The contract has been endorsed by several Republicans, like House Republican Leader Kevin McCarthy (CA-23) and Representative Dan Crenshaw (TX-02).[277]

The impact of young conservatives calling in politicians on the right is evident in the growth of the Conservative Climate Caucus—the name is self-explanatory—which, with seventy-three members, is now one of the largest Congressional caucuses.[278]

Change across party lines is perhaps most notably happening on the state level. Four out of the five states that have done the most to reduce emissions in the last few years through the passage of innovative bipartisan policy have Republican governors: Larry Hogan of Maryland, Brian Kemp of Georgia, Mike Dunleavy of Alaska, and Chris Sununu of New Hampshire.[279] Of the twenty-four states in the US that have specific executive and or statutory greenhouse gas emissions targets, six are governed by a Republican.[280] In addition, red states like Texas, Oklahoma, Florida, and Kansas are among the country's leaders in renewable energy production.[281]

Even some of the most staunch conservatives, like Florida Governor Ron DeSantis, are expanding electric vehicle infrastructure and working on expanding renewables into low-income areas.[282]

I remain hopeful that climate action can become a bipartisan and, in fact, a nonpartisan problem through the ongoing efforts of groups like ACC and citizens who regularly engage in conversations about the climate crisis with people who do not vote the same way as they do.

We are doing the hard work of depoliticizing climate change every time we push ourselves to engage in hard conversations with our neighbors, loved ones, friends, and family.

RECAP

Movements are hard to define but easy to identify. They exist everywhere and are behind every policy shift, rhetorical adjustment, and cultural change. Though the impact of social movements is incredibly difficult to pin down, the importance of movements cannot be understated. Acting as an impetus for change, movements enable progress, convince people of the importance of a problem, and provide a gateway for them to get involved in solving said problem.

Social movements also possess the ability to influence corporations and the private sector. The #PayUp campaign led by Fashion Revolution that calls on fashion brands to compensate garment workers for orders placed during COVID-19 that were canceled is a prime example of that. I would know; I've played a part in shaming many-a-brand on social media.

The power that can be yielded by voting with the dollar is what we will focus on next.

CHAPTER 10

CHANGE IN THE MARKETS

—

Like many other impassioned activists, I had a stage of life where I thought that money was antithetical to impact. Undeniably, this was an immensely privileged position to take and more symbolic rather than affecting, considering my parents still provided for me then. But then I got into the real world and realized everything is undergirded by a need for money, a need for sustenance—whether it is a nonprofit's continuous struggle to get funding or a hedge fund's need to give financial returns to its investors.

At this time, I simultaneously became frustrated by the pace of change that is often curtailed by red tape in the governmental and NGO spaces. I soon realized that, sometimes, positive change could be made in the most unlikely places. The writing of this chapter has only driven home that idea.

Twelve-year-old me would be incredulous to hear that I would later in life become the CEO of a retail and climate tech startup. But twenty-four-year-old me truly believes that the markets/private sector can be a high-impact way to make change.

The dollar is powerful. We can argue over the ethicality of that reality, but our climate time budget requires us not to fight against it, but with it.

And since some of the world's most prominent economists and business leaders now agree that planetary decline is the greatest threat to the economy, we are fighting less of an uphill battle.

Thankfully, there are many inroads to environmental progress within markets. When doing better is not only the right thing to do, but the economically prudent thing to do, it becomes a no-brainer.

The power of using money to make change is exemplified through the financial activism waged by the world against South Africa during Apartheid, the period of legal segregation. In the late '80s, citizens and politicians rallied together to put financial pressure on South Africa through boycotts and sanctions, respectively.

Consumers across the world mobilized together to boycott the purchase of South African goods. Consumer boycotts are a way to exert a "social control of business" and have been used in the movement for unionization, the peace movement led by Gandhi in India, the Civil Rights movement, and the movement to liberate occupied Palestinian territories.[283]

Despite opposition from the heads of state in each respective country, both the United States Congress and the UK Parliament passed international sanctions on the apartheid state.[284] By withholding aid and trade from the state, the actions of these governments exemplify what conscious consumerism (at scale) can look like.

Using the markets to force positive change might be relatively new within the environmental movement, but it is certainly fast growing.

In traditional accounting, businesses are taught to prioritize the bottom line—profit, return on investment, and shareholder value. But in recent years, businesses and NGOs have begun to prioritize what is called the triple bottom line—a business framework that challenges the profit-centric purpose of businesses enshrined in bottom line calculations.

The triple bottom line prioritizes the "three Ps": profit, planet, and people. In doing so, it reframes the very purpose of businesses.[285]

The trend toward the triple bottom line is not mere conjecture. In 2019, 181 chief executive officers of America's biggest corporations, part of the Business Roundtable, collectively overturned the twenty-two-year-old policy statement which defined a corporation's purpose as maximizing shareholder returns and replaced it with an amended policy statement that values sustainable practices and giving back to communities.[286]

As unfortunate as it is, power and money are inextricably linked. Throughout history, money has often been used to enhance and accumulate power in the hands of the few. But turning this pattern on its head and linking powerful change with money can yield an immense impact.

Institutions that have traditionally been leveraged to hoard wealth for stakeholders now see the value of community shareholders. Entities are learning that there will be no business on a dead planet, and that a fundamental responsibility of business is to create value without destroying ecological value. Talking about climate change, the CEO of BlackRock—the world's largest asset manager—has written that "awareness is rapidly changing, and I believe we are on the edge of a fundamental reshaping of finance."[287]

Let's walk through a few trends I've observed in the markets that have the potential to challenge the powers that have perpetuated climate change, transition us to a green grid, and fund a better future.

CLEAN ENERGY INNOVATION

In the Ted Talk called "The Case for Optimism on Climate Change,"[288] former US Vice President Al Gore outlines the promise that declining clean energy costs has for our fight against the climate crisis.

Renewable energy is becoming more affordable and widespread. Wind and solar plants have become 70 percent and 89 percent cheaper, respectively, in the last decade.[289] In 2020, 62 percent of total renewable power generation added was less expensive than any new production of the cheapest form of fossil fuels (coal).[290] In addition, renewable energy sources were the only types of energy sources for which demand increased during the pandemic.

Batteries are central to expanding the reach of clean energy even further. Innovations with battery technology have led to huge strides in the adoption of electric vehicles and renewable energy storage.

In the past decade, the average cost of lithium-ion batteries has declined by 89 percent, bringing down the cost of producing an electric car.[291] Forty years ago, the fact that General Motors would transition away from combustion engines by 2035 would be hearsay. Because of the competitive costs and improved performance of lithium-ion batteries, it is now the reality.[292]

Despite the supply chain struggles the industry faced brought on by COVID-19, battery storage has increased by over 1,200 percent from 2016 to 2021,[293] allowing for more clean power to be distributed with more efficient charges.

These advancements, in conjunction with renewable energy-friendly policies, have changed the landscape of our energy system. Between 2022 and 2023, fifty-one gigawatts of new solar and battery storage projects are expected to be added to the US power grid, making up 60 percent of new generating capacity.[294]

With these projects comes a huge potential for job creation, especially in areas of the world that, for decades, have existed as havens for auto manufacturing and coal mining. Despite the widespread layoffs during the pandemic, worldwide employment in renewable energy during the pandemic increased from 11.5 million in 2019 to 12 million in 2020.[295]

In the previous chapter, I highlighted the emergent trend of traditionally red (conservative) states switching to a more renewable-heavy energy grid. Of the states that added the most clean energy capacity to their grid in the first quarter of 2022, 60 percent of them have a majority Republican state Congress.[296]

Both states like California and those like Texas have made immense progress in the adoption of renewables. In other words, y'all, renewables are so promising that they are breaking that red/blue divide that has had a chokehold on the US for far too long.

Upending the entire energy infrastructure is a gargantuan task—but it's not impossible. Recent research that assesses energy systems in 145 countries shows that a full transition to clean energy across these countries—99.7 percent of the world's carbon dioxide emissions—would cost $61.5 trillion, most of which goes toward the deployment of technologies rather than innovation (95 percent of technologies needed for this decarbonization plan are already commercial).

This cost could be recovered in less than six years due to long-term energy and social-cost savings.[297]

Many people, myself included, got introduced to climate optimism through the inspiring journey of renewable energy in the last few decades. And although I believe that it is far from the only reason to be hopeful for our planet, its success and impact point to the truth that a reimagined future lies ahead.

CONSCIOUS CONSUMPTION

Metal straws alone will certainly not dig us out of the climate crisis. The same could be said with sustainable/ethical fashion, zero-waste groceries, etc. But the shift to more intentional, conscious consumerism should not be undercounted.

As consumers, we can move the needle—or the demand curve—simply by making (or not making) purchases and encouraging others to do the same. The needle functions more like a domino, given that engagement in a green behavior likely increases a consumer's engagement in another green behavior.[298] In other words, as an environmentally conscious consumer, you are probably not only concerned about buying sustainable clothes but also more likely to care about disposing of them sustainably.[299]

Now more than ever, you can be empowered to vote consciously with your dollar, using your purchasing power to drive industry-wide sustainability changes. A study published by Accenture, a consulting firm, revealed that 67 percent of consumers reported making more environmentally friendly and ethical purchases since the beginning of the pandemic.[300] Go us.

Although voting typically happens at most every two years, we vote with our dollars every day—and thus can engage in consumer advocacy. Kathryn Kellogg, an expert in low-waste living, has noticed a marked change in consumer advocacy in the seven years that she has used her social media platforms to share sustainability education.

It is not just that consumer advocacy is more widespread today than in previous years. It is also different than it ever was, as we are increasingly well-versed in greenwashing—deceptive marketing tactics or strategies that companies often use to make consumers think they are more sustainable than they are. As a result, we as consumers are more aware of root issues that inhibit companies from being sustainable and are better prepared to push back on marketing tactics.

Pre-2020, consumers seemed to fixate energy on avoiding unsustainable packaging—think images of turtles choking on plastic straws—but Kellogg notes that in the time since, she has noticed a shift in pressure toward companies' overall emissions and not just packaging.

Beyond the survey data showing a healthy rise in conscious consumerism, the trove of evidence supporting this trend lies in the changes made by brands also noticing this shift in consumer behavior. Big companies that were certainly not founded on principles of sustainability, like Walmart and even Amazon, have started quite impressive initiatives to minimize excess packaging, incentivize secondhand items, and reach net-zero emissions. It's likely that these companies—that have an outsized impact on the climate crisis—are not taking action on sustainability because of a sudden change in heart. The pressure of customers and shareholders powers this change.

Though these initiatives are not enough to right the wrongs that most corporations have invoked on our home planet, they are certainly something.

Many forces inhibiting us from environmental progress can seem insurmountable to challenge as one consumer. The movement of conscious consumerism, though, is very much within our grasp. We have the power to buy less and buy better. We can shift the demand curve in favor of better business practices that align with the future we hope to see.

ECOSYSTEM VALUATION

Much of the decimation of natural ecosystems and creatures has occurred because humans have not grasped their true value. Under much of Western history, nonhuman living things have been valued for how they can serve humans: how sheep can provide luxurious fur-lined jackets for a winter night out in the Big Apple; how tusks are grossly extracted from flailing, suffering elephant bodies to make ornate decor; or how beaver goo (castoreum) can be used to enhance vanilla and strawberry ice cream.

This mindset has caused what scientists called the Sixth Extinction, the first extinction primarily driven by human activity.

But what if that way of valuation was fundamentally altered?

Inspired by the biodiversity crisis and pleas from environmental scientists, Ralph Chami—an assistant director at the International Monetary Fund and financial economist specializing in fragile states and low and middle-income countries—decided to find a way to value ecosystems for more than meets the eye.

Central to this work is the idea that living things are worth more alive than dead, damaged, or destroyed.

Critically, the valuation falls not on the sale of the ecosystem or anything it contains, but rather the services it provides from being preserved, tailoring to the capitalist framework deeply embedded in most economies. Stewards of a given ecosystem do not have to face the moral quandary of whether to sell or destroy the ecosystem to make money and can instead continue to protect and steward it while being compensated.

Chami spoke passionately about the opportunities that ecosystem valuation has to uplift Indigenous communities who have, for centuries, been protecting nonhuman life. He told me about island nations sitting on around 20 percent of the world's seagrass. According to his calculations, the preservation of seagrass is valued at over a trillion dollars.

This new schema of assessing the value of nature is revolutionary, to say the least. It has gained attraction from some of the most highbrow financial executives who see the investment sense in preserving biodiversity.

The work doesn't stop at valuing the assets these ecosystems and creations provide; it continues through building markets where countries can sell certificates, much like carbon markets.

> **Carbon markets:** Trading system through which countries may buy or sell units of greenhouse gas emissions to meet their national limits on emissions.

Chami, alongside other leaders in the movement, has found an alternative to oil—a regenerative, diverse, and non-pollutive resource that has the potential to build up the economies of entire countries. For this reason, we can see the beginning of how ecosystem valuation will accrue massive benefits to both human and nonhuman life.

INSTITUTIONAL DIVESTMENT

The divestment movement is a natural extension of the anti-fossil fuels movement we touched on earlier. The movement calls for institutions to remove their investments in fossil fuel companies and reinvest in more ethical and stable industries.

It is estimated that 91.1 percent of oil and natural gas stocks are owned by institutions, investment advisors, and retirement funds.[301] This makes institutional divestment so powerful—it threatens to strip extractive companies of their power and driving force, which emanates from money.

There are a few immediate impacts of divestment. Regarding fossil fuel companies, divestment is a major loss of investment, large cash flows, loans, credit facilities, and insurance that they rely upon to fund projects.[302] Identifying sites, drilling wells, and building pipelines are extremely capital-intensive processes, requiring near constant access to funds that divestment seeks to threaten.

Divestment implies reinvestment. Institutions can reinvest their once-dirty money into companies paving the way for a cleaner, greener future.

Traction in this movement has grown immensely in the last decade, with some of the biggest academic institutions—Harvard University and Oxford University—and foundations—the Ford Foundation and the MacArthur Foundation—committing to divestment. Amy Gray, the senior climate finance strategist at Stand Earth, describes her experience seeing the divestment movement spread to "watching dominoes fall."[303]

The Divestment Database estimates that, as of publishing, 1,508 institutions—worth a combined $40.43 trillion—have divested from fossil fuels.[304] Yep, that's a trillion with a T.

A majority of these institutions are faith-based organizations, followed by educational institutions and philanthropic foundations. Pressure to divest largely comes from shareholders and members of an institution. So bravo to all who have had a hand in turning on the heat to create this change.

In October 2021, ABP, one of the world's largest pension funds, announced the sale of fifteen billion pounds worth of holdings in fossil fuel companies, citing "insufficient opportunity for us as a shareholder to push for the necessary, significant acceleration of the energy transition at these companies."[305]

After selling stock in over twenty coal companies, New York state's pension fund—which supports over one million New York state and municipal workers—announced that it would be divesting fossil fuel stocks in the next few years, and more generally from all climate change-perpetrating companies by 2040.[306] California currently has a divestment bill in the state Senate, which if passed would divest $9.9 billion of two of the largest pension funds in the world.[307]

Municipalities in the United States—like Seattle, New York City, and San Francisco—have also been stepping up to pass resolutions that ban pension funds from being invested in fossil fuels. Nice job, guys. I see you.

Institutional Divestment, to me, is one of the most impactful strategies for stopping the biggest perpetrators of the climate crisis. Every person is a part of an institution, whether a religious group, a university, or a workplace. Because of this, institutional divestment gives everyone a chance to pressure the institutions they are a part of. And as we know, putting on the heat can cause radical transformation.

SHAREHOLDER ACTIVISM

James Peck, the white Civil Rights activist perhaps most widely known for fearlessness in the Freedom Rides, took on a lesser-known approach in advocating for desegregation. In addition to sitting side-by-side with his Black friends, much to the chagrin of segregationists who set the bus on fire, he decided to invest in Greyhound and file a shareholder proposal with the company entitled "A Recommendation that Management Consider the Advisability of Abolishing the Segregated Seating System in the South," in hopes that it could move the company to desegregate seating.

A host of examples of shareholder proposals are used throughout history as a tool for urging powerful companies to, at least, recognize an important social issue and, at most, take action toward mitigating the issue.[308] Even if a vote for a resolution is not favorable, the process of the vote itself requires the company to make a statement on the resolution, providing important validity to the matter.

A 2019 survey of 439 institutional investors based in Europe revealed that a majority believe climate risk reporting is as important as financial reporting, and one-third believe that climate risk reporting is even more important than financial reporting.[309] These are my people.

In 2016, Follow This, an organization with the goal of pushing oil majors on climate action through shareholder resolutions, reached the necessary stake in Shell to allow them to introduce a shareholder resolution. That year, the resolution—which, in short, called on the company to "stop the exploration and acquisition of more oil and gas"[310]—was only supported by 2.7 percent of shareholders. Five years later, the organization's updated resolution achieved a landmark 30 percent shareholder support.[311] That amounts to a 1011 percent increase in the number of pro-climate old white men!

There was a time when shareholder resolutions were not an effective way to make a change within a company due to the power that boards and management had. Those days are over. Until the 1980s only two shareholder proposals opposed by management had ever achieved a majority vote. In the last year, seventy-four shareholder proposals achieved a majority vote—nineteen of which did not have management support.[312]

As of March 2022, one-fifth of 567 shareholder resolutions shared in the proxy season had to do with the climate,[313] signaling a push by climate-conscious shareholders that want to keep the companies they are investing in accountable.

Although the incremental change in climate finance can be discouraging, we must remember that this strategy is long-term. Despite the reality that a majority of shareholder votes on these resolutions are typically still in opposition to resolutions, shareholders are showing up to support resolutions in record numbers. Double digits are certainly better than single, and the trajectory of support signals a promising future for support for these resolutions.

In my opinion, the rise in environmental shareholder resolutions and receptivity to them is encouraging for one main reason that has the power to challenge our entire notion of corporate responsibility. Corporations regularly argue that their sole purpose is to maximize profits for their shareholders. But when shareholders bring non-finance-related issues to the annual general meetings, they are showing the companies that they care about more than financial returns. In tandem with the regulatory landscape pushing ESG reporting, it's clear people with the big bucks (and big power) are beginning to move the needle on climate finance.

PART III

CLIMATE OPTIMISM AROUND THE WORLD

NOTE

—

The last section was admittedly more Western-centric than I'd like. My perspective on climate solutions has largely been limited to what I've seen and researched in the United States, but that is far from the expansive range of solutions being pioneered all over the world.

This section highlights some of those solutions being spearheaded by changemakers who have firsthand seen the devastation of the climate crisis impact their homes, despite contributing little to nothing to the crisis.

I believe that the best solutions to the climate crisis come from people who know how to live harmoniously with the land and are experts in resilience and community care.

Their potential to make a significant difference has been highlighted in several ways, but I must emphasize that there is much I had to leave out in these stories. The range of stories included here is a crumb of all that exists out there. I implore you to go out and find more of these encouraging projects and people online and support their work.

CHAPTER 11

SHE4EARTH— SOUTH AFRICA

Shelot Masithi, a climate activist based in South Africa, talked to me about her experience starting She4Earth, an environmental education organization.

Shelot's childhood home sat at the foot of a mountain in her village. Growing up surrounded by nature, she was taken aback by the treeless vacant land around her university. After getting involved with environmental psychology research, she ascended into more international climate spaces and organizations. In these spaces, she was quickly confronted with the startling under-representation of people of African descent. Tokenism and exclusion were simultaneously rampant, and she often found herself in spaces where people were neglecting to talk about—or worse yet—dismissing the racialized context of climate change. She wanted to see people with her skin color brought to environmental conferences and organizations to speak up on their experiences, the inherent racial nature of climate change, and why climate justice necessitates racial justice and realized that for this to happen, she had to bring environmental education to classrooms across her country.

The remaining wounds of Apartheid are clear in the segregation of students between public and private schools.

During the Bantu education segregation system in the 1950s, non-white students went to poorly funded public schools under the control of the national government, which positioned children for labor-oriented blue-collar jobs, while white students in private schools got the chance to attain high levels of education.

Today, Shelot notes, South African public schools are marked by crumbling infrastructure and are unsurprisingly mostly Black, while private schools are mostly white.

As a part of the country's Black majority, Shelot saw education as an important avenue for change, especially for African youth, to "prepare them for the world that's coming."

When she shared her work with me for this book, she mentioned the disconnect between conversations about a Net-Zero world that go on in international meetings and marketing lingo and those that go on in classrooms. She asks me, "What is a Net-Zero world when you [ask] a child in primary school, especially in rural communities?" reflecting on the absence of climate change education in classrooms, especially rural ones.

> **Net-Zero World:** The balance established when the amount of greenhouse gases we add to the atmosphere is no more than that which is removed from the atmosphere.

The struggle students faced was not just a lack of climate change education but also a lack of practical knowledge.

Shelot describes the material taught in public schools as theory-based, whereas the material in private schools is far more practical.

When combined with the all-too-common phenomenon of young girls dropping out of school because of school fees, menstruation, and the responsibilities of taking care of other family members, the gaps in education became glaring to her. Shelot became motivated to "change that dimension of the education system...from theory to practical, because we need more practical [knowledge] today than ever before, with climate mitigation and climate change and everything happening all at once."

The educational program She4Earth has put together is grounded in the principles of Ubuntu—an African philosophy that prioritizes love, peace, kindness, respect, inclusivity, togetherness, wholeness, and compassion. For primary school children and disabled students, programs center on eco-mapping, a practice of familiarizing oneself with native plants and animals.

Although Shelot came in to fill a gap she had identified within the South African education system, she recognizes the need to work with the government for more resources and standardization of climate change education. Working on proposals for the Department of Education and the Department of Environmental Affairs, she hopes to provide accredited certifications or qualifications students can use for careers in the green economy.

She4Earth teaching students outdoors, Shelot Masithi

Zahra (Z): What is your advice for anyone who wants to mobilize others together for climate action in the face of devastation?

Shelot (S): There's quite a need for an individual to just sit with themselves and listen and let out [their] thoughts, emotions, and feelings.

With mental illnesses colliding with each other and climate change and poverty and inflation and everything, emotions and feelings can really be devastating, but you have to move on to jump your emotions into action.

So it's really important to let yourself feel what is going on within you. And I think for me, that you know, my anxiety has been the greatest source of my inspiration because a lot of ideas have come through in my anxious states, and I think I've been grateful for that.

Z: How do you think we can reimagine a more just and equal world and work toward it?

S: Bring all the people that you think and see that are underrepresented in these spaces, and give them a space. I mean, there's that concept of, you know, pass the mic, I think. So do that.

There's a lot of preaching that "we need more inclusion...we need more young people from the Global South represented." Stop talking about it and start doing it. Just give someone a space, an opportunity, even if it's five minutes or ten minutes. That could be the life-changing point or the point that changes the momentum for that person and for their community and for us as the global society of young people because we are the ones that are carrying the world.

If we don't learn from the disintegration of our leaders or our elders, and we continue with that disintegration then we are not going to achieve anything in the climate movement or climate action. We are fighting toward something that we cannot compromise. This is one common goal. So we need to come to that common ground and just learn.

I mean, get out of our egos and just be one for the first time in our lives.

HOW TO SUPPORT

The She4Earth team is seeking assistance from policy professionals to help them propose a partnership with the Department of Education.

Email **she4earth@gmail.com** to engage with She4Earth.

CHAPTER 12

PROJECT EIM— PHILIPPINES

A traverse across the Indian Ocean brings us to one of the 7,641 islands of the Philippines, Samar, where Vhon Michael Tobes lives.

Vhon grew up with the ocean on one side and the mountains on the other and as part of a tribe called Manaragat (People of the Sea). Knowledge of climate change was limited to the lived experiences of his community, witnessing sea level rises and resource limitations. But the reasoning behind the intensifying typhoons and erratic weather was far from conclusive, with many believing it was a consequence of spiritual intervention or a broken relationship with nature.

He first heard the term "climate change" in the eighth grade when it was glazed over in his science curriculum. This moment led him on a fateful path of discovery, and he eventually started an environmental educational organization called Project EIM (Educate, Inspire, Mobilize).

The organization develops easy-to-understand educational materials targeted at schoolchildren. The materials, which span a wide array of mediums, are centered on climate change education and climate change experience, intending to infuse scientific climate education with firsthand experience.

Meeting a need that hasn't been met with other organizations, Project EIM has translated the material into over thirteen languages and counting.

The need for this information is clear. Many people across the islands now know what climate change is but remain stuck in misinformation about why it is caused and the actual severity of the issue, amplified by the government's censorship.

Over the last two administrations, the government has acted illiberally, engaging in political censorship and disinformation campaigns and shutting down critical telecommunications infrastructure. Their aversion to climate action is no different. Climate action requires updated education, criticizes industry, and necessitates change in power structure, all things which are directly against the interest of the dictatorship.

In Vhon's words, "The people in power are creating and weaponizing myths and machinations to portray us (people who are advocating for not just climate education, but also education in general) because they're afraid of education. So they're portraying us as terrorists; they are portraying us so that people hate us because the government painted us as this type of monster that people should be wary of."

This political environment makes the Philippines one of the deadliest countries for environmental defenders like Vhon.

A climate change class during "Science Month," Vhon Michael Tobes

Zahra (Z): What is your advice for anyone who wants to mobilize others together for climate action in the face of devastation?

Vhon (V): Crying helps you. So let it all out. Then talk to your friends or literally just go on a walk with nature. See the beauty and feel. So you can cry all you want, you can go on trips, only one. Talk to your friends. It will heal you even while doing the work. Seeing the faces of the people you work with. Even though you all look tired, just seeing the resistance. You're seeing the bond that is connecting all of you. Just seeing that love for your country and love for your planet will heal you.

Activism or advocacy in general, is a constant process of learning and unlearning. So in order for me to actually learn things, and to move forward with my advocacy, I have to unlearn those toxic traits. Once we've finally given ourselves the basic knowledge and information about a certain issue that we are passionate about, we can start engaging with the people around us.

We can start with our family. We can start with our friends or our schoolmates. I would like to emphasize that advocacy means going beyond conversations with people who already agree with you. They shouldn't be your target. Target more people who have different opinions. And when you approach them, you shouldn't approach them with a facade that you're going to persuade them. Just engage in conversations. You might learn something from them.

Z: How do you think that we can reimagine a more just and equal world and work toward it?

V: We must take on and take down systems of oppression. These systems of oppression aren't only living or existing in the richest parts of the world. They can also exist [within] us, like the patriarchy. Patriarchy is a system that corrupts everyone, or it corrupts most of the people. It's not just men. So we must see that in every aspect and every corner of the world.

We must take on and take down, discuss how should we bring, how should we highlight and acknowledge those not just the entirety of the systems but also how they manifest those things. We should recognize them and then create solutions and do the opposite of what build it might be with might have been built by love for our that than or love for centralization. Then we decentralize. We do the opposite to bring it down.

HOW TO SUPPORT

- ◊ The Project EIM team needs help with translating content to languages.
- ◊ They are seeking out a publishing company to produce resources.
- ◊ Looking to produce short educational videos about the climate crisis.

Email **projecteimclimate@gmail.com** to get in touch and **projecteimcontact@gmail.com** to donate. Visit **@projecteim** on social media for more information.

CHAPTER 13

ECONOMIES OF PEACE— PAKISTAN

—

My mother country, Pakistan, has blistering summers that trump that of my hometown, Houston. They are tough for anyone to adjust to, even those who have lived there their whole life.

One summer, nine years ago, Fahad Rizwan began his climate journey by planting shady date trees along roads for people like himself going to and from work. Soon after, he received help from the military and local governments and joined hands with them to expand roadside planting and generate urban forests.

With the success of these local efforts and at the advice of grantmakers, Fahad decided to couple previous work with planting and increasing biodiversity with peace-building efforts. Pakistan is vulnerable to the overwhelming presence of terrorism and climate change—two ostensibly separate threats that have more to do with each other than one might think. As access to essential resources shrinks (food, water), violence increases. Studies even show the link between hotter temperatures and acts of aggression/crime.[314]

The region on the Pakistan/Afghanistan border called Kurram Agency (part of the Federally Administered Tribal Areas) is exceptionally vulnerable to both, with decades-long tensions between Shia and Sunni Muslim religious communities and agrarian stress exacerbated by the climate crisis.

Fahad's passion for the environment and people translated into the creation of GreenSquad Pakistan, an organization assisting Indigenous, tribal, and rural communities with adapting to the adverse effects of climate change. GreenSquad is uniquely fighting the joint struggle of violent conflict and climate conflict through their project called Economies of Peace.

The project involves training farmers in resilient and more environmentally friendly farming techniques while providing them with deradicalization training. Eventually, Fahad hopes, these local communities will rebuild their tensions through efforts to protect their shared environment.

Young boys in Kurram Agency, Fahad Rizwan

Zahra (Z): What is your advice for anyone who wants to mobilize others together for climate action in the face of devastation?

Fahad (F): Dismantle the very system because its very existence is based on profit motive. The contemporary economic structures are based on profit motive. It's so capital and money-oriented that you can't, you know, bring that view of nature and the human on the forefront.

So, the problems with the current economic structure, I think, they need to be addressed. The profit motive must be, you know, vanished from at least in these kinds of ventures of those ventures, you know, which are for climate action. And then there must be, I think, a huge and serious transformation of our whole production process. So, if the profit margins decline, let them decline and if there is no profit margin and you will adjust, you know, meeting your cost, run that production process in that way, because it's nature and humanity, which is facing the consequences of this profit-driven economic kind of production process, which is the very reason of this ecological crisis.

Z: How do you think that we can reimagine a more just and equal world and work toward it?

F: If you won't get yourself organized into a collective, the system will crush you. So you must build your own worker-owned cooperatives and you must build your own production base.

HOW TO SUPPORT

- ◊ The GreenSquad team is seeking assistance in the development of a mobile application.
- ◊ They are fundraising for microgrants for female farmers within the Economies of Peace program.

CHAPTER 14

ECO-LUSION— KENYA

The debate we have in the West and other rich nations about selecting our energy sources is a conversation that many people in rural communities in the Global South do not even imagine.

When I talked to Edikite, a renewable energy expert and the founder of the environmental organization Eco-lusion, he told me about the issues surrounding energy access in his community and how it impacts public health.

Edikite was born in the Kisumu County of Kenya and moved around quite a bit in his youth, observing the different ecosystems and environments around him.

In class seven, Edikite lost his stepmom. The year after, he lost his father and his mother. After investigating the causes of death, he found that all of his guardians passed from Tuberculosis—a disease that can in part be caused by air pollution exposure. He learned about the linkage between firewood as a cooking source, air pollution, and the disease that killed his guardians, and decided he had to dedicate his work to this.

According to government statistics, over 70 percent of people in Nyando (a sub-county within Kisumu where Edikite works) live below the poverty line, on less than a dollar a day. As such, they cannot afford minimum basic needs like food, water, and electricity.

Edikite told me about the high prices of propane gas driving village households to use kerosene, a gas that burns dirtier than propane. For cooking, most villagers use dried firewood or charcoal, both of which create a lot of pollution, causing health issues like respiratory diseases.

Through Eco-lusion, Edikite and his partner Jack are developing the technology for bioethanol stoves and, in the meantime, helping households transition to bamboo fuel and away from firewood/charcoal.

At the time of writing, they have planted up to 2,700 bamboo trees and are looking for partnerships and funding to meet the target of 25,000 seedlings by the end of 2022.

Women learning about briquetting process for clean cooking, Edikite Ochieng' Otieno

Zahra (Z): What is your advice for anyone who wants to mobilize others together for climate action in the face of diversity?

Jack: [You] need to come in with dedication and with a clear purpose and knowing what [you] are coming into and what [you] are fighting for. I'd advise you to stand still, push, and work with those who are willing to work with you.

Z: What are your words of advice for people who are advocating for change in a system that is unfriendly to them?

Edikite: I know in this journey you're going to face challenges; you are going to have to face resistance, even from within and from the government, but I think there are safe spaces. There are a lot of places where you can be an activist. You can use media; you can use social media.

It's so hard to organize the government, mostly in Africa, because we pretend to be so democratic as per se, but it is not still there. But there are a lot of international partnerships you can work with. Through partnership, you can air out your problem to a partner in another country with less resistance and somebody can maybe bring this thing to light in a way that the situation you are in will be highlighted.

HOW TO SUPPORT

The Eco-lusion team is fundraising for new machine that carbonates sugarcane to biochar to help make household briquettes.

Email **ochiengedikite@gmail.com** for more information.

CHAPTER 15

PLANTING HOPE— PALESTINE

—

When I picture a refugee camp, I don't know exactly what I picture but it certainly is not a "jungle of cement," as Hamdi Hato described the Al-Am'ari refugee camp in which he was raised until the age of twelve.

As a young child growing up in Al-Am'ari, he recalls the lack of a sewage network, infrastructure, electricity, and space and an overabundance of concrete walls and huddled bodies. The camp has a population density higher than Mumbai, India, and a remarkable lack of greenery.

While living in Al-Am'ari, Hamdi saw greenery only when his family traversed outside the camp. But he was overwhelmed with the beauty and healing powers of greenery during his years at university when he traveled around the world as part of an international study visit and was inspired by the roof gardens in London and Kuala Lumpur.

A decade later, Hamdi returned to the Al-Am'ari refugee camp with the mission to greenify roofs, pave a way for food sovereignty, and provide women with life-changing jobs.

Due to the Israeli occupation, most Palestinians depend on Israeli food systems and cannot cultivate food at scale for their own communities.

Through Planting Seeds, Hamdi has worked with sixty-seven young women to establish fifty green rooftops throughout the camp, each providing jobs, food security, and opportunities to thrive.

Rooftop garden in Al-Am'ari camp, Hamdi Hato

Zahra (Z): What is your advice for anyone who wants to bring others together for climate action in the face of devastation?

Hamdi (H): Bringing people together for climate action requires a human-centric approach, as is similar to many other forms of cooperation. There are countless roles to play as an individual exercising climate resilience and the necessary adaptations vary in every place that you may go. Increased public awareness and genuine community engagement is a starting point to inform more people about what they can do to contribute to a healthy environment on a global, national, regional, or local scale. This information can inspire passion and it is the job of leaders and organizers to make it possible to turn this passion into action and impact. Every community has access to differing resources, has varying ideologies, and faces different types of challenges. Adaptability is important in order to maximize engagement and enthusiasm of contributors under a wide range of environmental conditions.

Z: How do you think that we can reimagine a more just and equal world and work toward it? What is your advice for that?

H: As mentioned, this project is happening in Al-Am'ari refugee camp, meaning the participants have already had refugee status since their birth. In addition to being political refugees, climate change threatens to impose further challenges in the lives of Palestinians. This project is a form of resilience to the effects of climate change and ineffective natural resource management, establishing greater agency among the women of Al-Am'ari regarding the management of a self-sufficient food supply system as well as a community atmosphere. It is said that Palestinian refugees used to sleep with their feet outside of the tents to symbolize that they would not adapt their self-esteem or dignity to that of the tent, rather they would work to build a "tent" that reflected their inner potential and met their standards of living, despite having to start from virtually nothing. The women of this project are refusing to become refugees twice over, this time not due to displacement, but in the face of climate change or food insecurity.

With the objective of a more just and equal world, it is important that our voices become more amplified, and we continue to value education in our communities. Despite the numerous challenges faced by Palestinian society, the population remains ambitious, imaginative, resourceful, and most of all hopeful that internal and external efforts can lead to an end to the social and political injustices that hamper our desires to maximize our potential as individuals and as a nation.

HOW TO SUPPORT

- ◊ The Planting Hope initiative is fundraising for expansion of Rooftop Gardening in Al-Am'ari.
- ◊ With financial backing, they will be able to provide stipends to women working in the camp.

Email **Hatuhamdi@gmail.com** for information on how to get involved.

PART IV

YOUR ROLE

CHAPTER 16

WHAT YOU CAN DO

—

Amongst the headlines, scientific reports, and policy papers, the answer to "What if we don't act?" is resoundingly loud. The proclamation that the earth *could* be inhabitable in a decade too often sends people like you and me into a state of shock and distress, while politicians and power-hungry people find ways to hoard wealth before the climate clock nears its end.

I advocate that to make the changes that this planet so desperately needs, we must reframe the question to "What if we do?" This question should be posed in every sphere of impact, to people already involved in the climate movement and those who have yet to be enticed.

Within the courts, we can ask ourselves and others: *What if we begin to enshrine ecosystems with the right to exist and thrive without human interruption?* How might that change our relationship with the living beings around us?

In the streets, we can ask ourselves and others: *What if we organize another nonpartisan movement like Earth Day 1970, when millions of people organized for environmental policy?* How might we build bridges, heal divides across political lines, and effectively pass policies that protect?

In engaging with the markets, we can ask ourselves and others: *What if we leverage the massive investment potential within green tech to decarbonize our grid?* How might we transform our dependency on harmful fossil fuels while creating jobs for millions and investing in marginalized communities?

The stakes are high, but that means that the potential, too, is high.

—

After all the information and inspiration we've experienced together, I'd like to end the book the way it started—with practical tips for practicing climate optimism. By this point, you should be inspired to take action and hopeful about the action getting off the ground all over the world.

It's nice to know about all the good things going on, but it's even better to be a part of the sweeping movement to protect our planet and all that inhabit it.

I'm not claiming that these encouraging trends are a panacea to the climate crisis. But I believe that to move forward at all, we must stop waiting around for a silver bullet or a miracle.

In writing this book, I have found several new initiatives I hope to get more involved in within the climate change movement.

I often say that the best thing *you* can do for the planet sits at the intersection of your privileges, positionality, and passions. Dr. Ayana Elizabeth Johnson, a marine biologist and founder of the Urban Ocean Lab, offers another way of finding your climate superpower by urging people to ask themselves:

◊ **"What brings you joy?"**

◊ **"What are you good at?"**

◊ **"What work needs doing?"**

Both frameworks have been helpful for me in selecting which sphere I want to be a part of. Given my privilege and positionality of having a large social platform and passion for moving the needle on climate action through policy, I knew what I wanted and needed to do was invest time in bipartisan coalition building. I'm excited to join in existing efforts and work to build stronger bridges.

Everyone has an essential role to play in this movement. Becoming an active participant in climate action does not require you to forego your job, your responsibilities, or truckloads of money.

—

As long as there is grass that grows and water that flows, I believe that the earth is worth fighting for.

As long as there are people from different viewpoints engaging in recursive discussions and debates, I believe that we are capable of working toward progress.

As long as there are young people, I believe there is a moral imperative to fight for their futures.

Hope is one of the few renewable resources that we have.

So as long as we are here on this earth, we must continue sowing seeds of hope, tending to the fertile ground that birthed us, sustains us, and that will be our resting place.

ACKNOWLEDGMENTS

—

To all the powerful voices I interviewed for this book: Ted Hamilton, Jojo Mehta, Amy Westervelt, Amy Gray, Katie Surma, Benji Backer, Jojo Mehta, Andrew Behar, Kathryn Kellogg, Mike Meno, Ralph Chami—all of whom are fervent stewards of this earth—thank you for your time and your work. Moriah Lavey, thank you for your words that dance and for allowing me to feature them here. And to Shelot Masithi, Hamdi Hato, Fahad Rizwan, Edikite Ochieng' Otieno, Jack Steve Otieno, and Vhon Michael Tobes, your passion for our shared home is beyond inspiring. Thank you for all that you have done and all that you will continue to do.

ABOUT THE AUTHOR

—

Zahra is a youth climate activist and entrepreneur that shares action items and tidbits of hopeful news for the planet to over 100K people on her social media platforms. Since graduating university in 2021, she has consulted on climate communications and sustainability for Google, Youtube, and Meta, published a book, and given a TED talk. Most recently, Zahra became the CEO and founder of In the Loop, a circular economy software startup on a mission to make secondhand selling and purchasing as easy as retail.

REFERENCES

INTRODUCTION

1. Lieberman, Charlotte. "Why We Romanticize the Past." *New York Times*, April 2, 2021. https://www.nytimes.com/2021/04/02/smarter-living/why-we-romanticize-the-past.html.

2. Olivia Eubanks. "'Unprecedented' Named People's Choice 2020 Word of the Year by Dictionary.Com." ABC News, December 16, 2020. https://abcnews.go.com/Politics/unprecedented-named-peoples-choice-2020-word-year-dictionary/story?id=74735664.

3. Major, Claire. "Unprecedented Times and Innovation." *Innovative Higher Education* 45, no. 6 (October 23, 2020): 435-36. https://doi.org/10.1007/s10755-020-09528-4.

4. World Bank Group. "Decline of Global Extreme Poverty Continues but Has Slowed: World Bank." https://www.worldbank.org/en/news/press-release/2018/09/19/decline-of-global-extreme-poverty-continues-but-has-slowed-world-bank, September 19, 2018. The World Bank.

5. World Health Organization. "Global Health Estimates: Life Expectancy and Healthy Life Expectancy," 2019. https://www.who.int/data/gho/data/themes/mortality-and-global-health-estimates/ghe-life-expectancy-and-healthy-life-expectancy#:~:text=Globally%2C%20life%20expectancy%20has%20increased,reduced%20years%20lived%20with%20disability.

6. Roser, Max, and Esteban Ortiz-Ospina. "Literacy." Our World in Data, September 20, 2018. https://ourworldindata.org/literacy#citation.

7. World Health Organization. "Infant Mortality," 2018. https://www.who.int/data/gho/data/themes/topics/indicator-groups/indicator-group-details/GHO/infant-mortality#:~:text=Globally%2C%20the%20infant%20mortality%20rate,to%204.0%20million%20in%202018.

8. Local Burden of Disease WaSH Collaborators. "Mapping Geographical Inequalities in Access to Drinking Water and Sanitation Facilities in Low-Income and Middle-Income Countries, 2000-17." *The Lancet Global Health* 8, no. 9 (September 2020): 1162-85. https://doi.org/10.1016/S2214-109X(20)30278-3.

9. Ritchie, Hannah, Max Roser, and Pablo Rosado. "Access to Energy." Our World in Data, 2022. https://ourworldindata.org/energy-access.

10. Ritchie, Hannah, and Max Roser. "Air Pollution." Our World in Data, January 2021. https://ourworldindata.org/air-pollution.

11. DeSilver, Drew. "Despite Global Concerns about Democracy, More than Half of Countries Are Democratic." Pew Research Center, May 14, 2019. https://www.pewresearch.org/fact-tank/2019/05/14/more-than-half-of-countries-are-democratic/.

12. Ida, Tetsuji. "Climate Refugees—the World's Forgotten Victims." World Economic Forum, June 18, 2021. https://www.weforum.org/agenda/2021/06/climate-refugees-the-world-s-forgotten-victims/.

13. Price, Shepard. "Climate Change, Global Warming Could Cost the World $178 Trillion in Coming Decades per Deloitte." Alton *Telegraph*, May 23, 2022. https://www.thetelegraph.com/news/article/Climate-change-could-cost-the-world-178-trillion-17191978.php.

14. Sommer, Lauren. "Climate Change Is The Greatest Threat To Public Health, Top Medical Journals Warn." *NPR*, September 7, 2021. https://www.npr.org/2021/09/07/1034670549/climate-change-is-the-greatest-threat-to-public-health-top-medical-journals-warn.

15. Femia, Francesco, and Caitlin E. Werrell. "Climate Change as Threat Multiplier: Understanding the Broader Nature of the Risk." The Center for Climate and Security, February 12, 2015.

16. Lorenz, Taylor. "The News Is Making People Anxious. You'll Never Believe What They're Reading Instead." *New York Times*, April 14, 2020. https://www.nytimes.com/2020/04/14/style/good-news-coronavirus.html.

17. Cecchi-Dimeglio, Paola. "Why Sharing Good News Matters." *MIT Sloan Management Review*, June 17, 2020. https://sloanreview.mit.edu/article/why-sharing-good-news-matters/.

18. World Health Organization. "Mental Health and Psychosocial Considerations during the COVID-19 Outbreak." World Health Organization, March 18, 2020. https://www.who.int/docs/default-source/coronaviruse/mental-health-considerations.pdf.

CHAPTER 1: INFORMATION OVERLOAD

19. Cocirillo, Francesco. "Cirillo Consulting GmbH: Services, Products, Software to Enhance Your Productivity." Cirillo Company. Accessed June 29, 2022. https://francescocirillo.com/.

20   Shammas, Michael. "Why a Simple Time-Management System Can Revolutionize How You Work–and Live." SSRN, December 11, 2019. https://papers.ssrn.com/sol3/papers.cfm?abstract_id=3492102.

21   Bulao, Jacquelyn. "How Much Data Is Created Every Day in 2022?" Techjury, June 3, 2022. https://techjury.net/blog/how-much-data-is-created-every-day/#gref.

22   Insider Intelligence Editors. "US Adults Added 1 Hour of Digital Time in 2020." Insider Intelligence, January 26, 2021. https://www.emarketer.com/content/us-adults-added-1-hour-of-digital-time-2020?ecid=NL1001.

23   Bulao, Jacquelyn. "How Much Data Is Created Every Day in 2022?"

24   Kemp, Simon. "Digital 2021: The Latest Insights into the 'State of Digital.'" We Are Social UK. *We Are Social*, February 1, 2022. https://wearesocial.com/uk/blog/2021/01/digital-2021-the-latest-insights-into-the-state-of-digital/.

25   Dictionary.com. "Disinformation Definition & Meaning." Accessed June 29, 2022. https://www.dictionary.com/browse/disinformation.

26   Dictionary.com. "Misinformation Definition & Meaning." Accessed June 29, 2022. https://www.dictionary.com/browse/misinformation.

27   Mohsin, Maryam. "10 TikTok Statistics You Need to Know." *Oberlo*, July 10, 2022. https://www.oberlo.com/blog/tiktok-statistics.

28   Nagata, Jason M., Catherine A. Cortez, et al. "Screen Time Use Among US Adolescents During the COVID-19 Pandemic Findings From the Adolescent Brain Cognitive Development (ABCD) Study." *JAMA Pediatrics* 176, no. 1 (November 1, 2021): 94–96. https://doi.org/10.1001/jamapediatrics.2021.4334.

29   Silverman, Craig. "This Analysis Shows How Viral Fake Election News Stories Outperformed Real News On Facebook." *BuzzFeed News*, November 16, 2016. https://www.buzzfeednews.com/article/craigsilverman/viral-fake-election-news-outperformed-real-news-on-facebook#.ei12Ya7XZ.

30   Boland, Brian. "Organic Reach on Facebook." Meta for Business. https://www.facebook.com/business/news/Organic-Reach-on-Facebook.

31   Walker, Mason, and Katerina Eva Matsa. "News Consumption Across Social Media in 2021." Pew Research Center, September 20, 2021. https://www.pewresearch.org/journalism/2021/09/20/news-consumption-across-social-media-in-2021/.

CHAPTER 2: NEGATIVITY BIAS

32   IATA, 2021.

33 Painter, James. *Climate Change in the Media: Reporting Risk and Uncertainty.* I.B. Tauris & Co. Ltd in association with the Reuters Institute for the Study of Journalism, University of Oxford, 2013. https://reutersinstitute.politics.ox.ac.uk/sites/default/files/research/files/Climate%2520Change%2520in%2520the%2520Media.pdf.

34 Trussler, Marc, and Stuart Soroka. "Consumer Demand for Cynical and Negative News Frames." *The International Journal of Press/Politics* 19, no. 3 (July 2014): 360-79. https://doi.org/10.1177/1940161214524832.

35 Peeters, Guido, and Janusz Czapinski. "Positive-Negative Asymmetry in Evaluations: The Distinction Between Affective and Informational Negativity Effects." *European Review of Social Psychology* 1, no. 1 (1990): 33-60. https://doi.org/10.1080/14792779108401856.

36 Vaish, Amrisha, Tobias Grossmann, and Amanda Woodward. "Not All Emotions Are Created Equal: The Negativity Bias in Social-Emotional Development." *Psychological Bulletin* 134, no. 3 (May 1, 2008). https://doi.org/10.1037/0033-2909.134.3.383.

37 Soroka, Stuart, and Stephen McAdams. "News, Politics, and Negativity." *Political Communication* 32, no. 1 (February 3, 2015): 1-22. https://doi.org/https://doi.org/10.1080/10584609.2014.881942.

38 Hilbig, Benjamin E. "Good Things Don't Come Easy (to Mind)." *Experimental Psychology* 59, no. 1 (January 1, 2011): 38-46. https://doi.org/https://doi.org/10.1027/1618-3169/a000124.

39 Soroka, Stuart, and Stephen McAdams. "News, Politics, and Negativity," February 3, 2015.

CHAPTER 3: THE PRIVILEGE GAP

40 Velho, Astride, and Oscar Thomas-Olalde. "Othering and Its Effects: Exploring the Concept." Writing Postcolonial Histories of Intercultural Education, January 1, 2011. https://www.academia.edu/42889355/Othering_and_its_effects_exploring_the_concept?auto=citations&from=cover_page.

41 Matthews, Dylan. "23 Charts and Maps That Show the World Is Getting Much, Much Better." *Vox*, November 24, 2014. https://www.vox.com/2014/11/24/7272929/global-poverty-health-crime-literacy-good-news.

42 The World Bank. "Poverty Headcount Ratio at $2.15 a Day (2017 PPP) (% of Population)." *The World Bank*, 2021. https://data.worldbank.org/indicator/SI.POV.DDAY.

43   Heglar, Mary Annaïse. "Climate Change Isn't the First Existential Threat." ZORA, February 18, 2020. https://zora.medium.com/sorry-yall-but-climate-change-ain-t-the-first-existential-threat-b3c999267aa0.

CHAPTER 4: FAILURE OF IMAGINATION

44   National Commission on Terrorist Attacks upon the United States. "The 9/11 Commission Report," July 22, 2004. https://www.govinfo.gov/content/pkg/GPO-911REPORT/pdf/GPO-911REPORT-24.pdf.

CHAPTER 5: ECHO CHAMBERS

45   Menczer, Filippo, and Thomas Hills. "Information Overload Helps Fake News Spread, and Social Media Knows It." *Scientific American*, December 1, 2020. https://www.scientificamerican.com/article/information-overload-helps-fake-news-spread-and-social-media-knows-it/.

46   Jagiello, Robert D., and Thomas T. Hills. "Bad News Has Wings: Dread Risk Mediates Social Amplification in Risk Communication." *Risk Analysis* 38, no. 10 (May 29, 2018): 2193-2207. https://doi.org/10.1111/risa.13117.

CHAPTER 6: THINGS AREN'T ALWAYS AS THEY SEEM

47   Biabani, Zahra. Twitter post. April 4, 2022, 6:44 p.m. https://twitter.com/ZahraNurBiabani/status/1511006870055661585.

48   Miller, Bruce. "Sulfur Oxides Formation and Control." *Fossil Fuel Emissions Control Technologies*, 197-242. Butterworth-Heinemann, 2015. https://www.sciencedirect.com/science/article/pii/B978012801566700004X.

49   Schmalensee, Richard, and Robert N. Stavins. "The SO2 Allowance Trading System: The Ironic History of a Grand Policy Experiment–American Economic Association." *Journal of Economic Perspectives* 27, no. 1 (2013): 103-22. https://doi.org/10.1257/jep.27.1.103.

50   Environmental Defense Fund. "How Economics Solved Acid Rain." September 2018. https://www.edf.org/approach/markets/acid-rain.

51   World Meteorological Organization. "Executive Summary: Scientific Assessment of Ozone Depletion: 2018." World Meteorological Organization, Global Ozone Research and Monitoring Project–Report No. 58, 2018.

52 Young, Paul J., Anna B. Harper, et al. "The Montreal Protocol Protects the Terrestrial Carbon Sink." *Nature* 596, no. 7872 (August 19, 2021). https://doi.org/http://dx.doi.org/10.1038/s41586-021-03737-3.

53 Hartnett White, Kathleen, and Brent Bennett. "The US Leads the World in Clean Air: The Case for Environmental Optimism." Texas Public Policy Foundation, April 2019. https://www.texaspolicy.com/wp-content/uploads/2018/11/2019-04-RR-US-Leads-the-World-in-Clean-Air-ACEE-White.pdf.

54 United States Environmental Protection Agency. "Highlights from the Clean Air Act 40th Anniversary." August 12, 2021. https://www.epa.gov/clean-air-act-overview/highlights-clean-air-act-40th-anniversary.

55 United States Environmental Protection Agency. "Progress Cleaning the Air and Improving People's Health." March 9, 2022. https://www.epa.gov/clean-air-act-overview/progress-cleaning-air-and-improving-peoples-health#pollution.

56 Lewis, Michelle. "EGEB: 76% of Proposed Coal Plants Have Been Canceled since 2015." *Electrek*, September 14, 2021. https://electrek.co/2021/09/14/egeb-76-of-proposed-coal-plants-have-been-canceled-since-2015/.

57 Wayland, Michael. "Auto Executives Say More than Half of US Car Sales Will Be EVs by 2030, KPMG Survey Shows." *CNBC*, November 30, 2021. https://www.cnbc.com/2021/11/30/auto-executives-say-more-than-half-of-us-car-sales-will-be-evs-by-2030-kpmg-survey-shows.html.

58 Markolf, Sam, Inês M.L. Azevedo, et al. "Pledges and Progress: Steps toward Greenhouse Gas Emissions Reductions in the 100 Largest Cities across the United States." *Brookings*, October 21, 2020. https://www.brookings.edu/research/pledges-and-progress-steps-toward-greenhouse-gas-emissions-reductions-in-the-100-largest-cities-across-the-united-states/#:~:text=One%20place%20to%20start%20such,GHG%20inventories%20and%20reduction%20targets.

59 E2. "After Hard Year, Promise of Unparalleled Jobs Growth," April 19, 2021. https://e2.org/reports/clean-jobs-america-2021/.

60 Chandler, David. "Explaining the Plummeting Cost of Solar Power." *MIT News*, November 20, 2018. https://news.mit.edu/2018/explaining-dropping-solar-cost-1120.

61 Teixeira, Ruy. "Americans Still Support Environmental Protection." Center for American Progress, May 14, 2012. https://www.americanprogress.org/article/public-opinion-snapshot-americans-still-support-environmental-protection/.

CHAPTER 7: LET'S TALK ABOUT EARTH WINS

62. Baazil, Diederik, and Laura Hurst. "Shell Loses Dutch Court Case Over Nigeria Oil Spills." *Bloomberg*, January 29, 2021. https://www.bloomberg.com/news/articles/2021-01-29/dutch-court-orders-shell-nigeria-to-compensate-for-oil-spills.

63. Willsher, Kim. "Court Convicts French State for Failure to Address Climate Crisis." *The Guardian*, February 3, 2021. https://www.theguardian.com/environment/2021/feb/03/court-convicts-french-state-for-failure-to-address-climate-crisis.

64. Piette, Betsey. "Fracking Banned in Delaware River Basin." *Workers World*, March 2, 2021. https://www.workers.org/2021/03/54802/.

65. Scherer, Glenn. "Landmark Decision: Brazil Supreme Court Sides with Indigenous Land Rights." Mongabay Environmental News, April 13, 2021. https://news.mongabay.com/2021/04/landmark-decision-brazil-supreme-court-sides-with-indigenous-land-rights/#:~:text=Landmark%20decision%3A%20Brazil%20Supreme%20Court%20sides%20with%20Indigenous%20land%20rights,-by%20Ana%20Ionova&text=Brazil%27s%20Supreme%20Federal%20Court%20.

66. Williams, Jessica. "Green by 2050: New Orleans City Council Orders Entergy to Cut Emissions." NOLA.com, May 20, 2021. https://www.nola.com/news/business/article_5297cdc4-b982-11eb-903e-b3ae5b66d433.html.

67. Boffey, Daniel. "Court Orders Royal Dutch Shell to Cut Carbon Emissions by 45% by 2030." *The Guardian*, May 26, 2021. https://www.theguardian.com/business/2021/may/26/court-orders-royal-dutch-shell-to-cut-carbon-emissions-by-45-by-2030.

68. Ajmera, Ankit. "Developer Officially Cancels Keystone XL Pipeline Project Blocked by Biden." *Reuters*, June 10, 2021. https://www.reuters.com/business/energy/tc-energy-terminates-keystone-xl-pipeline-project-2021-06-09/.

69. WGME Staff. "Maine Becomes First State to Pass Law to Divest from Fossil Fuels." *WGME*, June 22, 2021. https://wgme.com/news/local/maine-becomes-first-state-to-pass-law-to-spanest-from-fossil-fuels.

70. Kaieteur News. "Oil Companies Ordered to Help Cover US$7.2 B Gulf of Mexico Cleanup Bill," July 9, 2021. https://www.kaieteurnewsonline.com/2021/07/09/oil-companies-ordered-to-help-cover-us7-2-b-gulf-of-mexico-cleanup-bill/.

71. Madeson, Frances. "Activists Are Using the 'Climate Necessity Defense' in Court—and Winning." *Truthout*, December 24, 2021. https://truthout.org/articles/activists-are-using-the-climate-necessity-defense-in-court-and-winning/.

72. Sebag, Gaspard. "French State Fined For Failing to Clean Up Nation's Air." *Bloomberg*, August 4, 2021. https://www.bloomberg.com/news/articles/2021-08-04/french-state-fined-for-failing-to-clean-up-nations-air#:~:text=The%20French%20state%20was%20criticized,several%20parts%20of%20the%20nation.

73. The Associated Press. "A Federal Judge Has Thrown Out Approvals For A Major Oil Project In Alaska." *NPR*, August 19, 2021. https://www.npr.org/2021/08/19/1029223070/federal-judge-thrown-out-approvals-oil-project-alaska-conocophillips#:~:text=U.S.%20District%20Court%20Judge%20Sharon,defended%20the%20project%20in%20court.

74. Armao, Mark. "Montana Tribe Finalizes Historic $1.9 Billion Water Rights Settlement." *Grist*, September 24, 2021. https://grist.org/agriculture/indigenous-water-rights/.

75. Simmons, Matt. "B.C. Pays Blueberry River First Nations $65 Million as 195 Projects Approved before Court Victory Proceed." *The Narwhal*, October 8, 2021. https://thenarwhal.ca/bc-blueberry-river-agreement/.

76. Newsroom. "Landmark Decision Gives Legal Teeth to Protect Environmental Defenders." *Modern Diplomacy*, October 24, 2021. https://moderndiplomacy.eu/2021/10/24/landmark-decision-gives-legal-teeth-to-protect-environmental-defenders/.

77. Villeneuve, Marina. "New York Voters Approve Right To A Clean Environment." *The Post-Journal*, November 3, 2021. https://www.post-journal.com/news/latest-news/2021/11/new-york-voters-approve-right-to-a-clean-environment-2/.

78. Yale Environment 360. "Landmark Ruling Blocks Mining in Ecuadorian Forest, Citing Rights of Nature." Yale School of the Environment, December 3, 2021. https://e360.yale.edu/digest/landmark-ruling-blocks-mining-in-ecuadorian-forest-citing-rights-of-nature.

79. Ibid.

80. Phillips, Anna, and Joselow, Maxine. "Judge Throws out Massive Gulf of Mexico Oil and Gas Lease Sale." *Washington Post*, January 28, 2022. https://www.washingtonpost.com/climate-environment/2022/01/27/biden-gulf-of-mexico-lease-sale/.

81. Brown, Kimberly. "Ecuador's Top Court Rules for Stronger Land Rights for Indigenous Communities." Mongabay Environmental News, February 9, 2022. https://news.mongabay.com/2022/02/ecuadors-top-court-rules-for-stronger-land-rights-for-indigenous-communities/.

82. Yale Environment 360. "Student Campaigners at Five Major Universities File Legal Complaints Against Fossil Fuel Investment." Yale School of the Environment, February 16, 2022. https://e360.yale.edu/digest/student-campaigns-at-yale-princeton-stanford-vanderbilt-and-mit-file-legal-complaints-calling-for-divestment.

83   Phillips, Anna. "Appellate Court Rules Biden Can Consider Climate Damage in Policymaking." *Washington Post*, March 16, 2022. https://www.washingtonpost.com/climate-environment/2022/03/16/social-cost-of-carbon-ruling/.

84   Jedra, Christina. "Honolulu Scores a Win against Big Oil in Climate Change Lawsuit." *Grist*, March 14, 2022. https://grist.org/article/honolulu-scores-a-win-against-big-oil-in-climate-change-lawsuit/.

85   Friends of the Earth. "Govt's Climate Strategy Deemed 'Unlawful' in Historic Ruling." *Friends of the Earth*, July 18, 2022. https://friendsoftheearth.uk/climate/govts-climate-strategy-deemed-unlawful-historic-ruling.

86   Shieber, Jonathan. "For the First Time the US DOT Is Carving out Budget for Climate and Environmental Justice Projects." *TechCrunch*, February 18, 2021. https://techcrunch.com/2021/02/17/for-the-first-time-the-us-dot-carving-out-budget-for-climate-and-environmental-justice-projects/.

87   InsiderNJ. "Jersey City Vertical Farming Program to Open in Two Public Housing Locations, Targeting Most Vulnerable Residents." *CLPHA*, March 10, 2021. https://clpha.org/news/2021/jersey-city-vertical-farming-program-open-two-public-housing-locations-targeting-most.

88   "Michael Regan Sworn 16th EPA Administrator." EPA. Environmental Protection Agency, March 11, 2021. https://www.epa.gov/newsreleases/michael-s-regan-sworn-16th-epa-administrator.

89   Hughes, Owen. "Right to Repair Moves Forward for Your Broken Devices. But Campaigners Want to Go Much Further." *TechRepublic*, March 2, 2021. https://www.techrepublic.com/article/right-to-repair-moves-forward-for-your-broken-devices-but-campaigners-want-to-go-much-further/.

90   The Library of Congress. "Australia: Ban on Single-Use Plastic Products Enacted in Australian Capital Territory." May 10, 2021. https://www.loc.gov/item/global-legal-monitor/2021-05-10/australia-ban-on-single-use-plastic-products-enacted-in-australian-capital-territory/.

91   Nexus Media News. "Brenda Mallory Confirmed To Head CEQ." April 15, 2021. https://nexusmedianews.com/top_story/brenda-mallory-confirmed-to-head-ceq/.

92   "Governor Baker Signs Climate Legislation to Reduce Greenhouse Gas Emissions, Protect Environmental Justice Communities." Commonwealth of Massachusetts, March 26, 2021. Governor's Press Office. https://www.mass.gov/news/governor-baker-signs-climate-legislation-to-reduce-greenhouse-gas-emissions-protect-environmental-justice-communities.

93   The Nature Conservancy in Washington. "Washington Takes Bold Action with Passage of Climate Commitment Act–The Nature Conservancy in Washington." *The Nature Conservancy in Washington*, April 24,

2021. https://www.washingtonnature.org/fieldnotes/2021/4/23/washington-takes-bold-action-with-passage-of-climate-commitment-act-legislature-olympia.

94   Anderson, Erik. "San Diego County Supervisors Add Environmental Justice Office." *KPBS Public Media*, May 19, 2021. https://www.kpbs.org/news/environment/2021/05/19/san-diego-county-considers-environmental-justice.

95   Marca Chile. "Chile Becomes the First Latin American Country to Ban Single-Use Plastics." *Marca Chile*, June 18, 2021. https://marcachile.cl/en/innovation-entrepreneurship/chile-becomes-the-first-latin-american-country-to-ban-single-use-plastics/#:~:text=2021%20%7C%20Innovation%20%26%20Entrepreneurship-,Chile%20becomes%20the%20first%20Latin%20American%20country%20to%20ban%20single,the%20end%20of%20the%20year.&text=According%20to%20a%20study%20carried,plastics%20per%20year%20in%20Chile.

96   Laville, Sandra. "Turkey to Ban Plastic Waste Imports." *The Guardian*, May 19, 2021. https://www.theguardian.com/world/2021/may/19/turkey-to-ban-plastic-waste-imports.

97   "Oil Companies Pull out of Alaska's Arctic National Wildlife Refuge." EcoWatch. EcoWatch, November 29, 2022. https://www.ecowatch.com/oil-companies-drilling-leases-arctic-national-wildlife-refuge.html.

98   *Maui Now*. "80 Acres on O'ahu Transferred from Federal Government to Native Hawaiian Community." June 14, 2021. https://mauinow.com/2021/06/14/80-acres-on-oahu-transferred-from-federal-government-to-native-hawaiian-community/.

99   Hillbrand, Alex, and Mona Yew. "China Accepts Kigali Amendment, Will Phase Down HFCs." *NRDC*, June 22, 2021. https://www.nrdc.org/experts/alex-hillbrand/china-accepts-kigali-amendment-will-phase-down-hfcs.

100   EPA Press Office. "EPA Announces $50 Million to Fund Environmental Justice Initiatives Under the American Rescue Plan." US EPA, June 25, 2021. https://www.epa.gov/newsreleases/epa-announces-50-million-fund-environmental-justice-initiatives-under-american-rescue.

101   "Culver City Council Passes Historic Ordinance to Phase Out Oil Drilling," June 18, 2021. https://www.sierraclub.org/press-releases/2021/06/culver-city-council-passes-historic-ordinance-phase-out-oil-drilling.

102   Kelley, Alexandra. "DOI Returns More than 18k Acres of Land to Native American Tribes." *The Hill*, June 23, 2021. https://thehill.com/changing-america/respect/equality/559902-doi-returns-over-18000-acres-of-land-to-native-american/.

103   Rosenberg, Lizzy. "What to Know About the Now-Canceled Byhalia Connection Pipeline." *Green Matters*, July 14, 2021. https://www.greenmatters.com/p/byhalia-connection-pipeline.

104 Cockburn, Harry. "Argentina Becomes First Country to Ban Open-Net Salmon Farming Due to Impact on Environment." *The Independent*, July 8, 2021. https://www.independent.co.uk/climate-change/news/argentina-salmon-farming-ban-environment-b1880503.html.

105 Deutsche Welle. "Greenland Stops Oil and Gas Exploration." July 16, 2021. https://www.dw.com/en/greenland-stops-oil-and-gas-exploration-climate-costs-too-high/a-58294024.

106 McGuire, Peter. "Maine's Landmark Recycling Reform Law Will Take Years to Implement." *Portland Press Herald*, June 4, 2022. https://www.pressherald.com/2022/06/04/maines-landmark-recycling-reform-law-will-take-years-to-implement/#:~:text=Maine%20passed%20the%20nation%27s%20first,covered%20under%20a%20similar%20program.

107 Samar, Azeem. "Karachi's Largest Public Hospital to Run on Solar Power." *Gulf News*, June 17, 2021. https://gulfnews.com/world/asia/pakistan/karachis-largest-public-hospital-to-run-on-solar-power-1.79995220.

108 The Canadian Press. "Feds Reach Nearly $8B Deal with First Nations over Clean Drinking Water Lawsuit." *Global News*, July 30, 2021. https://globalnews.ca/news/8074983/feds-drinking-water-lawsuit-first-nations/.

109 Reuters Staff. "Norway to Spend $1.2 Bln on Renewable Projects in Developing Nations." *Reuters*, July 7, 2021. https://www.reuters.com/article/norway-climate/norway-to-spend-1-2-bln-on-renewable-projects-in-developing-nations-idUSL5N2OI3H0.

110 Milman, Oliver. "Washington State County Is First in US to Ban New Fossil Fuel Infrastructure." *The Guardian*, July 28, 2021. https://www.theguardian.com/us-news/2021/jul/28/washington-state-whatcom-county-ban-fossil-fuel-infrastructure.

111 New York State Energy Research and Development Authority. "Governor Cuomo Announces $52.5 Million Available for Community Solar Projects That Support Underserved New Yorkers." *New York State*, July 20, 2021. https://www.nyserda.ny.gov/About/Newsroom/2021-Announcements/2021-07-20-Governor-Cuomo-Announces-52-5-Million-Available-for-Community-Solar-Projects-that-Support-Underserved-New-Yorkers.

112 Krol, Debra Utacia. "Tribes Cheer Withdrawal of 2 Little Colorado Hydropower Projects but Fear a 3rd." *AZCentral*, August 5, 2021. https://www.azcentral.com/story/news/local/arizona-environment/2021/08/05/2-little-colorado-river-dam-projects-have-been-withdrawn/5408667001/.

113 HHS Establishes Office of Climate Change and Health Equity. HHS, August 30, 2021. https://www.hhs.gov/about/news/2021/08/30/hhs-establishes-office-climate-change-and-health-equity.html.

114. Lombrana, Laura Millan. "Toxic Leaded Gasoline Production Ends as Last Refinery Shuts Down." *Bloomberg*, August 30, 2021. https://www.bloomberg.com/news/articles/2021-08-30/toxic-leaded-petrol-production-ends-as-last-refinery-shuts-down?leadSource=uverify%20wall.

115. Sherfinsk, David. "Get 'folks like Me' Engaged on Climate, Says Black Greenpeace USA Head." *Thomson Reuters Foundation News*. September 14, 2021. https://news.trust.org/item/20210914112749-h5wyl/.

116. Cosgrove, Jaclyn. "L.A. County Takes First Steps to End Urban Oil Drilling." *Los Angeles Times*, September 16, 2021. https://www.latimes.com/california/story/2021-09-16/l-a-county-takes-first-steps-to-end-oil-drilling.

117. IANS. "England to Be First Country to Require New Homes to Include EV Chargers." *Business Standard*, September 12, 2021. https://www.business-standard.com/article/international/england-to-be-first-country-to-require-new-homes-to-include-ev-chargers-121091200367_1.html.

118. Tepler, Benjamin. "Four National Parks in Australia Returned to Indigenous Nations." *Backpacker*, September 30, 2021. https://www.backpacker.com/news-and-events/news/australia-gives-back-national-parks-to-indigenous-tribes/.

119. Kochis, Lauri. "A Closer Look at Philadelphia's Plastic Bag Ban." *Temple Now*, November 9, 2021. https://news.temple.edu/news/2021-11-09/closer-look-philadelphia-s-plastic-bag-ban.

120. Rivard, Ry. "PennEast Calls It Quits after a Yearslong Fight to Build a Gas Pipeline." *POLITICO*, September 27, 2021. https://www.politico.com/states/new-jersey/story/2021/09/27/penneast-calls-it-quits-after-a-years-long-fight-to-build-a-gas-pipeline-1391442.

121. Spiegel, Anna. "Major DC Conservation Group Will Omit Racist 'Audubon' From Its Name." *Washingtonian*, October 25, 2021. https://www.washingtonian.com/2021/10/25/audubon-naturalist-society-will-change-its-name-citing-racist-namesake/.

122. The Wildlife Trusts. "Long-Awaited Environment Act Passed." November 10, 2021. https://www.wildlifetrusts.org/news/long-awaited-environment-act-passed.

123. Associated Press. "Native American Confirmed as Head of National Park Service." *CBS4 WTTV-TV*, November 20, 2021. https://cbs4indy.com/news/native-american-confirmed-as-head-of-national-park-service/.

124. The White House. "Justice40 Initiative—Environmental Justice." July 25, 2022. https://www.whitehouse.gov/environmentaljustice/justice40/.

125. Peoples Dispatch. "After a Year of Struggle by Farmers, Indian Government Forced to Withdraw Farm Laws." November 19, 2021. https://peoplesdispatch.org/2021/11/19/after-a-year-of-struggle-by-farmers-indian-government-forced-to-withdraw-farm-laws/.

126   Treisman, Rachel. "A California Redwood Forest Has Officially Been Returned to a Group of Native Tribes." *NPR*, January 26, 2022. https://www.npr.org/2022/01/26/1075778055/california-redwood-forest-native-american-tribes.

127   AFP. "Honduras Bans Open-Pit Mining." *Digital Journal*, March 1, 2022. https://www.digitaljournal.com/world/honduras-bans-open-pit-mining/article.

128   Yang, Maya. "California Plan Would Give $100m to Indigenous Leaders to Buy Ancestral Lands." *The Guardian*, March 19, 2022. https://www.theguardian.com/us-news/2022/mar/18/california-indigenous-tribes-purchase-land?campaign_id=49&emc=edit_ca_20220321&instance_id=56292&nl=california-today&regi_id=169072724&segment_id=86104&te=1&user_id=e79afafa76b45cffef614df73f889094.

129   Winters, Joseph. "Rappahannock Tribe Gets 465 Acres of Land Back on the Chesapeake Bay." *Grist*, April 4, 2022. https://grist.org/indigenous/rappahannock-tribe-gets-465-acres-land-back-chesapeake-bay/.

130   Armstrong, Martin. "Environmental Protection vs. Economic Growth." Statista, April 12, 2022. https://www.statista.com/chart/27234/us-survey-prioritize-environment-vs-economy/.

131   Beer, Mitchell. "Quebec Becomes World's First Jurisdiction to Ban Oil and Gas Exploration." The Energy Mix, April 13, 2022. https://www.theenergymix.com/2022/04/13/quebec-becomes-worlds-first-jurisdiction-to-ban-oil-and-gas-exploration/.

132   Uibu, Katri. "Tasmania Goes into Carbon Negative, with Researchers Saying Native Forests Must Be Preserved." *ABC News*, May 2, 2022. https://www.abc.net.au/news/2022-05-03/tas-carbon-negative-emission-levels-credited-to-stopping-logging/101032008.

133   Fossil Fuel Treaty. "London Endorses Global Treaty Calling for an End to the Fossil Fuel Age and a Just Energy Transition—The Fossil Fuel Non-Proliferation Treaty." The Fossil Fuel Non-Proliferation Treaty press release, June 28, 2022. https://fossilfueltreaty.org/london.

134   Winters, Joseph. "California Passes Nation's Toughest Plastic Reduction Bill." *Grist*, July 1, 2022. https://grist.org/regulation/california-new-legislation-fights-plastic-pollution/?utm_source=newsletter&utm_medium=email&utm_campaign=beacon.

135   Yale Environment 360. "Massachusetts Bill Would Allow Residents to Contribute to Climate-Vulnerable Countries When Filing Taxes." Yale School of the Environment, July 27, 2022. https://e360.yale.edu/digest/massachusetts-climate-finance-taxes.

136   Osborne, Margaret. "Hawaii Closes Its Last Coal-Fired Power Plant." *Smithsonian Magazine*, September 6, 2022. https://www.smithsonianmag.

com/smart-news/hawaii-closes-its-last-coal-fired-power-plant-180980707/.

137. Kuta, Sarah. "Hydrogen-Powered Passenger Trains Are Now Running in Germany." *Smithsonian Magazine*, September 7, 2022. https://www.smithsonianmag.com/smart-news/hydrogen-powered-passenger-trains-are-now-running-in-germany-180980706/.

138. Manghani, Ravi. "How Falling Costs Will Secure Solar's Dominance in Power." Wood Mackenzie, January 2021. https://www.woodmac.com/horizons/how-falling-costs-will-secure-solars-dominance-in-power/?utm_campaign=horizons.

139. DeSantis, Rachel. "Ikea Buys 11K Acres of Land in Georgia for Conservation." *People*, February 1, 2021. https://people.com/human-interest/ikea-buys-10000-acres-land-georgia-protect-it/#:~:text=From%20Being%20Developed-,Ikea%20Buys%20More%20Than%2010%2C000%20Acres%20of%20Land%20in,Protect%20It%20From%20Being%20Developed&text=Ikea%27s%20investment%20group%20recently%20acquired,plant%20and%20wildlife%20species%20safe.

140. Rainforest Action Network. "Citi Becomes First US Bank to Restrict Financing for Companies Expanding Coal Power," March 15, 2021. https://www.ran.org/press-releases/citi-becomes-first-u-s-bank-to-restrict-financing-for-companies-expanding-coal-power/.

141. Amelang, Sören. "Retired Coal Plant in Germany to Be Converted to Green Hydrogen Site." Clean Energy Wire, January 22, 2021. https://www.cleanenergywire.org/news/retired-coal-plant-germany-be-converted-green-hydrogen-site.

142. Redling, Adam. "Mattel Launches Toy Takeback Program in Support of Sustainability Efforts." *Recycling Today*, May 11, 2021. https://www.recyclingtoday.com/article/mattel-playback-toy-recycling-packaging-reuse/#:~:text=on%20May%2010-,Mattel%20PlayBack%20is%20a%20toy%20takeback%20program%20designed%20to%20recover,products%20and%20packaging%20by%202030.

143. Roselund, Christian. "Green Steel Is Picking up Steam in Europe." *Canary Media*, June 23, 2021. https://www.canarymedia.com/articles/clean-industry/green-steel-is-coming-sooner-than-you-think.

144. Guardian staff and agencies. "Harvard University Will Divest Its $42bn Endowment from All Fossil Fuels." *The Guardian*, September 10, 2021. https://www.theguardian.com/environment/2021/sep/10/harvard-university-spanest-endowment-fossil-fuels.

145. Dolan, Leah. "Luxury Fashion Giant Kering Bans Fur across All Its Businesses." *CNN*, September 24, 2021. https://edition.cnn.com/style/article/kering-bans-fur/index.html.

146   Kerber, Ross. "MacArthur Foundation Joins Investment Shift Away Fossil Fuels." *Reuters*, September 22, 2021. https://www.reuters.com/business/sustainable-business/macarthur-foundation-joins-investment-shift-away-fossil-fuels-2021-09-22/.

147   Peters, Jay. "Google and YouTube Will Cut off Ad Money for Climate Change Deniers." *The Verge*, October 7, 2021. https://www.theverge.com/2021/10/7/22715102/google-youtube-climate-change-deniers-ads-monetization.

148   Kalvapalle, Rahul. "U of T to Divest from Fossil Fuel Investments, Create Climate-Positive Campus." *University of Toronto News*, October 27, 2021. https://www.utoronto.ca/news/u-t-spanest-fossil-fuel-investments-create-climate-positive-campus.

149   Parker, Lisa, and Tom Jones. "A Win For Right To Repair: Apple Announces Future Repair Program for Popular Devices." *NBC Chicago*, December 15, 2021. https://www.nbcchicago.com/consumer/a-win-for-right-to-repair-apple-announces-future-repair-program-for-popular-devices/2708422/.

150   Reed, Jessica. "United Completes First Commercial Flight with 100% Sustainable Fuel." Avionics International, December 8, 2021. https://www.aviationtoday.com/2021/12/08/united-completes-first-commercial-flight-100-sustainable-fuel/#:~:text=United%20Completes%20First%20Commercial%20Flight%20with%20100%25%20Sustainable%20Fuel,-By%20Jessica%20Reed&text=United%20has%20achieved%20a%20historical,while%20using%20100%25%20sustainable%20fuel.

151   Haber, Taylor. "College Formally Announces Plan to Divest from Fossil Fuels." *The Dartmouth*, October 8, 2021. https://www.thedartmouth.com/article/2021/10/college-announces-spanestment-plans.

152   Corp., Green Hydrogen International. "Green Hydrogen International Announces Hydrogen City, Texas—The World's Largest Green Hydrogen Production and Storage Hub." PR Newswire, March 3, 2022. https://www.prnewswire.com/news-releases/green-hydrogen-international-announces-hydrogen-city-texas--the-worlds-largest-green-hydrogen-production-and-storage-hub-301494988.html.

153   Shivaram, Deepa. "Starbucks Plans to Phase out Paper Cups in the US and Canada." *NPR*, March 16, 2022. https://www.npr.org/2022/03/16/1086862986/starbucks-plans-to-phase-out-paper-cups-in-the-u-s-and-canada.

154   MacMillan, Douglas, and Maxine Joselow. "SEC Plans to Force Public Companies to Disclose Greenhouse Gas Emissions." *Washington Post*, March 15, 2022. https://www.washingtonpost.com/business/2022/03/15/sec-climate-emissions-rule/.

155   Colarossi, Jessica. "Boston University to Divest from Fossil Fuel Industry." *BU Today*, September 23, 2021. https://www.bu.edu/articles/2021/boston-university-spanest-from-fossil-fuel-industry/.

156 Clifford, Catherine. "Stripe Teams up with Major Tech Companies to Commit $925 Million toward Carbon Capture." *CNBC*, April 12, 2022. https://www.cnbc.com/2022/04/12/stripe-alphabet-meta-shopify-mckinsey-spur-carbon-capture-market.html.

157 Leber, Rebecca. "The Long-Shot Campaign to Get Big Banks out of Fossil Fuels." *Vox*, April 29, 2022. https://www.vox.com/23046282/banks-climate-shareholder-votes-fossil-fuels.

158 Wilson, Janet. "California Just Shy of 100% Powered by Renewables for First Time." *Palm Springs Desert Sun*, May 1, 2022. https://www.desertsun.com/story/news/environment/2022/05/01/california-100-percent-powered-renewables-first-time/9609975002/.

159 Chu, Jennifer. "Concept for a Hybrid-Electric Plane May Reduce Aviation's Air Pollution Problem." *MIT News*, January 14, 2021. https://news.mit.edu/2021/hybrid-electric-plane-pollution-0114.

160 United Nations Environment Programme. "Rewilding Sees Jaguars Return to Argentina's Wetlands." February 11, 2021. https://www.unep.org/news-and-stories/story/rewilding-sees-jaguars-return-argentinas-wetlands.

161 Tangermann, Victor. "This Is the World's First Home Hydrogen Battery." *Futurism*, January 23, 2021. https://futurism.com/the-byte/worlds-first-home-hydrogen-battery.

162 BBC News. "Solar Energy Empowers Young Women in Yemen," February 3, 2021. https://www.bbc.com/news/av/world-middle-east-55894840.

163 Visram, Talib. "Seville Is Turning Its Iconic Oranges into Electricity." *Fast Company*, March 2, 2021. https://www.fastcompany.com/90609626/seville-is-turning-its-iconic-oranges-into-electricity.

164 Sahagún, Louis. "Will the Wolf Survive? The Stunning California Odyssey of OR-93, a Wolf on a Mission." *Los Angeles Times*, March 25, 2021. https://www.latimes.com/environment/story/2021-03-25/stunning-california-odyssey-or93-wolf-on-a-mission.

165 Taylor, Derrick Bryson. "America's Bald Eagle Population Has Quadrupled." *New York Times*, March 25, 2021. https://www.nytimes.com/2021/03/25/climate/how-many-bald-eagles-united-states.html#:~:text=Researchers%20say%20the%20population%20is%20now%20above%20300%2C000.&text=The%20bald%20eagle%20population%20in,on%20the%20brink%20of%20extinction.

166 Ethirajan, Anbarasan. "Nepal Rhino Numbers Rise in 'exciting' Milestone." *BBC News*, April 13, 2021. https://www.bbc.com/news/world-asia-56731079.

167 Rosane, Olivia. "Giant Pandas No Longer Endangered Thanks to Conservation Efforts." *World Economic Forum*, July 14, 2021. https://www.weforum.org/agenda/2021/07/giant-pandas-endangered-species-conservation-efforts-china-animals/#:~:text=China%20no%20longer%20

considers%20giant%20pandas%20an%20endangered%20species%20in,are%20now%20considered%20%22vulnerable.%22.

168   Cannon, Jay. "They Pulled 63,000 Pounds of Trash from the Great Pacific Garbage Patch, but That's Just the Start." MSN, October 29, 2021. https://www.msn.com/en-us/weather/topstories/they-pulled-63-000-pounds-of-trash-from-the-great-pacific-garbage-patch-but-that-s-just-the-start/ar-AAQ5lc9.

169   Mongabay. "Cuba Boosts Marine Protected Coverage with New Area Spanning Reefs to Mangroves." *Mongabay Environmental News*, January 27, 2022. https://news.mongabay.com/2022/01/cuba-boosts-marine-protected-coverage-with-new-area-spanning-reefs-to-mangroves/.

170   Dreier, Frederick. "Doctors in Canada Can Now Prescribe a Trip to a National Park." *Backpacker*, February 3, 2022. https://www.backpacker.com/news-and-events/news/doctors-in-canada-can-now-prescribe-a-trip-to-a-national-park/.

171   WWF Report Highlights Tiger Population Gains for the Year of the Tiger." World Economic Forum, February 2022. https://www.weforum.org/agenda/2022/02/wwf-tiger-population-wildlife-lunar-new-year/#:~:text=The%20global%20tiger%20population%20is,contributed%20to%20the%20animal's%20recovery

172   Bush, Evan. "Coral Resiliency Offers Encouraging Signs despite Warming Oceans." *NBC News*, March 11, 2022. https://www.nbcnews.com/science/environment/coral-resiliency-offers-encouraging-signs-warming-oceans-rcna19502.

173   Ferris, Nick. "Scotland's Forests Are the Largest They Have Been for 900 Years." *New Statesman*, April 8, 2022. https://www.newstatesman.com/chart-of-the-day/2022/04/scotlands-forests-are-the-largest-they-have-been-for-900-years.

174   Cox, Lisa. "Scientists Report 'Heartening' 30% Reduction in Plastic Pollution on Australia's Coast." *The Guardian*, June 8, 2022. https://www.theguardian.com/australia-news/2022/jun/09/scientists-report-heartening-30-reduction-in-plastic-pollution-on-australias-coast#:~:text=Australia%20news-,Scientists%20report%20%27heartening%27%2030%25%20reduction%20in,plastic%20pollution%20on%20Australia%27s%20coast&text=The%20amount%20of%20plastic%20pollution,research%20by%20Australia%27s%20science%20agency.

175   NOAA Office of National Marine Sanctuaries. "Proposed Designation of Hudson Canyon National Marine Sanctuary." Office of National Marine Sanctuaries, 2022. https://sanctuaries.noaa.gov/hudson-canyon/.

176   Shanghai Metals Market News. "Turkey Discovers 694 Million Mt of Rare Earth Element Reserves, with Infrastructure Construction Starting This Year_SMM," July 5, 2022. https://news.metal.com/newscontent/101881567/emturkeyem-discovers-694-million-mt-

of-rare-earth-element-reserves-with-infrastructure-construction-starting-this-year?utm_campaign=%5Bcampaign_name%5D&utm_medium=email&utm_source=Sailthru&utm_term=Keep%20Cool#.YsTGcg0I6p0.twitter.

177 Archie, Ayana. "Coral Levels in Some Parts of the Great Barrier Reef Are at the Highest in 36 Years." *NPR*, August 4, 2022. https://www.npr.org/2022/08/04/1115539492/coral-great-barrier-reef-australia.

178 The Ocean Cleanup. "System 002," July 13, 2021. https://theoceancleanup.com/milestones/system-002/.

CHAPTER 8: CLIMATE OPTIMISM IN THE COURTROOM

179 Biabani, Zahra. "What Filing a Complaint against My University Taught Me." *Teen Vogue*, April 5, 2022. https://www.teenvogue.com/story/fossil-free-five-coalition-complaint.

180 Ibid.

181 Britannica, T. Editors of Encyclopedia. "Commonwealth v. Hunt." Encyclopedia Britannica, September 21, 2016. https://www.britannica.com/event/Commonwealth-v-Hunt.

182 Spitzer, Elianna. "What Is Judicial Activism?" ThoughtCo, June 22, 2020. https://www.thoughtco.com/judicial-activism-definition-examples-4172436#:~:text=of%20Judicial%20Activism-,Brown%20v.,Clause%20of%20the%2014th%20Amendment.

183 Setzer, Joana, and Catherine Higham. "Global Trends in Climate Change Litigation: 2021 Snapshot." *The Centre for Climate Change Economics and Policy and The Grantham Research Institute on Climate Change and the Environment*, July 2021. https://www.lse.ac.uk/granthaminstitute/wp-content/uploads/2021/07/Global-trends-in-climate-change-litigation_2021-snapshot.pdf.

184 Ibid.

185 Ibid.

186 Bouwer, Kim, and Joana Setzer. "Climate Litigation as Climate Activism: What Works?" *The British Academy*, 2020. https://www.thebritishacademy.ac.uk/publications/knowledge-frontiers-cop26-briefings-climate-litigation-climate-activism-what-works/.

187 Hamilton, Ted, March 24, 2022.

188 Leghari v. Federation of Pakistan. 25501 W.P. 201 (2015).

189 County of Maui v. Hawaii Wildlife Fund. 590 US (2020).

190   Earthjustice. "The Clean Water Case of the Century," October 21, 2021. https://earthjustice.org/features/supreme-court-maui-clean-water-case.

191   Sabin Center for Climate Change Law. "Milieudefensie et al. v. Royal Dutch Shell Plc.," April 5, 2019. http://climatecasechart.com/non-us-case/milieudefensie-et-al-v-royal-dutch-shell-plc/.

192   Gloucester Resources Limited v. Minister for Planning. NSWLEC 7; 234 LEGRA 257 (2017).

193   Environmental Law Australia. "Gloucester Resources ('Rocky Hill') Case." Accessed June 29, 2022. http://envlaw.com.au/gloucester-resources-case/.

194   Gbemre v. Shell Petroleum Development Company of Nigeria Ltd. et al. FHC/B/CS/53/05 (2005).

195   Torres-Spelliscy, Ciara. "The History of Corporate Personhood." Brennan Center for Justice, April 8, 2014. https://www.brennancenter.org/our-work/analysis-opinion/history-corporate-personhood.

196   Citizens United v. FEC. 558 US 310 (2010).

197   Bainbridge, Emma. "Indigenous Mobilization in Ecuador." The Emergence of CONAIE. Brown University Library. Accessed June 29, 2022. https://library.brown.edu/create/modernlatinamerica/chapters/chapter-6-the-andes/moments-in-andean-history/indigenous-mobilization-in-ecuador/.

198   Turner, Stephen J., Dinah L. Shelton, et al., eds. *Environmental Rights: The Development of Standards*. Cambridge: Cambridge University Press, 2019. Doi:10.1017/9781108612500.

199   *Politico*. "Pennsylvania Election Results 2016: President Live Map by County, Real-Time Voting Updates." Election Hub, December 13, 2016. https://www.politico.com/2016-election/results/map/president/pennsylvania/.

200   Community Environmental Legal Defense Fund. "Rights of Nature: Timeline." CELDF, June 3, 2021. https://celdf.org/rights-of-nature/timeline/.

201   "Before the Madurai Bench of Madras High Court," April 4, 2022. https://www.livelaw.in/pdf_upload/mother-nature-416320.pdf.

202   Urgenda Foundation v. The State of Netherlands, HAZA C/09/456689 (2015).

203   Ibid [4.97].

204   Sabin Center for Climate Change Law. "VZW Klimaatzaak v. Kingdom of Belgium & Others," August 19, 2016. http://climatecasechart.com/non-us-case/vzw-klimaatzaak-v-kingdom-of-belgium-et-al/.

205   Kusnetz, Nicholas, Inside Climate News, Katie Surma, and Yuliya Talmazan. "'Ecocide' Movement Pushes for a New International Crime: Environmental Destruction." *NBC News*, April 7, 2021. https://www.nbcnews.com/news/world/ecocide-movement-pushes-new-international-crime-environmental-destruction-n1263142.

206 Stop Ecocide Foundation. "Independent Expert Panel for the Legal Definition of Ecocide." DocJax, June 2021. https://docjax.com/documents/ttpsstatic1squarespacecomstatic581f63c8414fb5367a5248f0t59e5f4babe42d60ae3debe4f1508242622017flyer-2018/.

207 Raftopoulos, Malayna, and Joanna Morley. "Ecocide in the Amazon: The Contested Politics of Environmental Rights in Brazil." *The International Journal of Human Rights* 24, no. 10 (November 8, 2019): 1616-41. https://doi.org/https://doi.org/10.1080/13642987.2020.1746648.

208 Office of the Prosecutor, *Policy Paper On Case Selection And Prioritisation* (n. 1) paras. 7.

209 Ecocide Law. "Existing Ecocide Laws." Accessed June 30, 2022. https://ecocidelaw.com/existing-ecocide-laws/.

210 Leghari v. Federation of Pakistan, 25501 W.P. 201 (2015).

211 Ibid.

212 Juliana v. United States (2015) Case No: 6:15-cv-01517-TC.

213 Harvard Law Review. "Juliana v. United States." *Harvard Law Review*, March 10, 2021. https://harvardlawreview.org/2021/03/juliana-v-united-states/.

214 "Order in Pending Case 18A65 United States, et al. v. USDC OR," July 30, 2018. https://static1.squarespace.com/static/571d109b04426270152febe0/t/5b5f6e246d2a734c55c26e57/1532980772367/18A65+United+States.+v.+USDC+OR+Order.pdf.

215 Peel, J., & Osofsky, H. (2018). "A Rights Turn in Climate Change Litigation?" *Transnational Environmental Law*, 7(1), 37-67. doi:10.1017/S2047102517000292.

216 Brulle, Robert J. "Networks of Opposition: A Structural Analysis of US Climate Change Countermovement Coalitions 1989-2015." *Sociological Inquiry* 91, no. 3 (October 21, 2019): 603-624. https://doi.org/10.1111/soin.12333.

217 Ibid.

218 Climate Files. "1965 API President 'Meeting the Challenges of 1966.'" November 8, 1965. https://www.climatefiles.com/trade-group/american-petroleum-institute/1965-api-president-meeting-the-challenges-of-1966/.

219 Supran, Geoffrey, and Naomi Oreskes. "Assessing ExxonMobil's Climate Change Communications (1977-2014)." *Environmental Research Letters* 12, no. 8 (August 23, 2017). https://doi.org/10.1088/1748-9326/aa815f.

220 ExxonMobil, "Unsettled Science," advertisement, *New York Times*, March 23, 2000.

221 Li, Mei, Gregory Trencher, and Jusen Asuka. "The Clean Energy Claims of BP, Chevron, ExxonMobil and Shell: A Mismatch between Discourse, Actions

and Investments." *PLOS ONE* 17, no. 2 (February 16, 2022). https://doi.org/10.1371/journal.pone.0263596.

222 N., Sönnichsen, and Statista. "Share of Gross Estimated Capital Expenditure on Low-Carbon Energy of Selected Oil Companies from 2018 to 2021*." *Statista*, July 15, 2021. https://www.statista.com/statistics/1112857/big-oil-clean-energy-capex-share-by-company/.

223 Hasemyer, David. "Maryland's Capital City Joins a Long Line of Litigants Seeking Climate-Related Damages from the Fossil Fuel Industry." *Inside Climate News*, February 24, 2021. https://insideclimatenews.org/news/24022021/annapolis-maryland-climate-oil-lawsuit/.

224 Press Release: City of Annapolis Announces Lawsuit to Hold 26 Fossil Fuel Defendants Accountable. (2021, February 23). *Annapolis.gov*. Retrieved from https://www.annapolis.gov/CivicAlerts.aspx?AID=1194.

225 Bruggers, James. "California Attorney General Investigates the Oil and Gas Industry's Role in Plastic Pollution, Subpoenas Exxon." *Inside Climate News*, April 30, 2022. https://insideclimatenews.org/news/30042022/california-attorney-general-exxon-plastic-pollution/.

226 "Commonwealth Bank Shareholders Sue over 'Inadequate' Disclosure of Climate Change Risks." 2017. *The Guardian*. August 8, 2017. https://www.theguardian.com/australia-news/2017/aug/08/commonwealth-bank-shareholders-sue-over-inadequate-disclosure-of-climate-change-risks.

227 Ramirez v. Exxon Mobil Corp, N.D. Tex. 3:16-cv-3111 (2016).

228 Law Students for Climate Accountability. "The 2021 Law Firm Climate Change Scorecard," August 2021. https://static1.squarespace.com/static/5f53fa556b708446acb4dcb5/t/611dba29c5ad3077663d4947/1629338162366/2021+Law+Firm+Climate+Change+Scorecard.pdf.

229 Federman, Adam. "Revealed: US Listed Climate Activist Group as 'Extremists' alongside Mass Killers." *The Guardian*, January 13, 2020. https://www.theguardian.com/environment/2020/jan/13/us-listed-climate-activist-group-extremists.

230 Hamilton, Ted. *Beyond Fossil Law: Climate, Courts, and the Fight for a Sustainable Future*. OR Books, 2018. https://www.orbooks.com/catalog/beyond-fossil-law/.

231 Scheidel, Arnim, Daniela Del Bene, et al. "Environmental Conflicts and Defenders: A Global Overview." *Global Environmental Change* 63 (July 2020): 102104. https://doi.org/10.1016/j.gloenvcha.2020.102104.

232 International Center for Not-For-Profit Law. "US Protest Law Tracker." Accessed June 30, 2022. https://www.icnl.org/usprotestlawtracker/?location=&status=enacted&issue=&date=&type=legislative#.

233   The Alabama Legislature. "Alabama Legislature." Alison on Alabama legislature. Accessed June 30, 2022. http://alisondb.legislature.state.al.us/Alison/default.aspx.

234   Office of the High Commissioner For Human Rights. "Mandates of the Special Rapporteur on the Promotion and Protection of the Right to Freedom of Opinion and Expression, and the Special Rapporteur on the Rights to Freedom of Peaceful Assembly and of Association." Perma.cc, March 27, 2017. https://perma.cc/4RMX-ZQ5A.

235   Massachusetts v. Gore. Boston Mun. Ct., Mass., No. 1606CR000923, March 27, 2018.

236   Tribunal de Grande Instance de Lyon, 19168000015, September 16, 2019.

237   Rainforest Foundation US. "Mass Trial of Indigenous Leaders to Take Place in Peru, May 14, 2014." Rainforest Fund, May 12, 2014. https://www.rainforestfund.org/mass-trial-of-indigenous-leaders-to-take-place-in-peru-may-14-2014/.

238   International Labour Organization. "Ratifications for Peru." Accessed June 30, 2022. https://www.ilo.org/dyn/normlex/en/f?p=NORMLEXPUB%3A11200%3A0%3A%3ANO%3A%3AP11200_COUNTRY_ID%3A102805.

239   Rights and Resources Initiative. "Five Years after the Bagua Protests, Indigenous Peoples Continue to Defend Their Right to Self-Determination." Rights + Resources, May 13, 2014. https://rightsandresources.org/blog/five-years-after-the-bagua-protests-indigenous-peoples-continue-to-defend-their-right-to-self-determination/.

240   Cultural Survival. "Peru: Indigenous Protestors at Bagua Declared Innocent of Police Deaths." *Cultural Survival*, October 24, 2016. https://www.culturalsurvival.org/news/peru-indigenous-protestors-bagua-declared-innocent-police-deaths.

241   Nosek, Grace. "The Climate Necessity Defense: Protecting Public Participation in the US Climate Policy Debate in a World of Shrinking Options." *Environmental Law* 49, no. 1 (2019): 249–61. https://www.jstor.org/stable/26794284.

242   Barritt, Emily, and Boitumelo Sediti. "The Symbolic Value of Leghari v. Federation of Pakistan: Climate Change Adjudication in the Global South." *King's College London Law School Research Paper, Forthcoming*, June 25, 2019.

CHAPTER 9: STARTING A MOVEMENT

243   Morris, Aldon. 1993. "Birmingham Confrontation Reconsidered." *American Sociological Review* 58: 5: 621–636.

244 Andrews, Kenneth T, and Sarah Gaby. "Local Protest and Federal Policy: The Impact of the Civil Rights Movement on the 1964 Civil Rights Act." Sociological Forum. Eastern Sociological Society, June 2015. https://sociology.unc.edu/wp-content/uploads/sites/165/2016/12/Gaby-Sociological-Forum.pdf.

245 International Center on Nonviolent Conflict. "The Success of Nonviolent Civil Resistance," November 4, 2013. https://www.nonviolent-conflict.org/resource/success-nonviolent-civil-resistance/.

246 Olzak, Susan, and Sarah A. Soule. "Cross-Cutting Influences of Environmental Protest and Legislation." *Social Forces* 88, no. 1 (September 1, 2009): 201-225. https://doi.org/10.1353/sof.0.0236.

247 Muñoz, John, Olzak, Susan, and Sarah A. Soule. "Going Green: Environmental Protest, Policy, and CO2 Emissions in US States, 1990-2007." *Sociological Forum*, 33, no. 2 (March 13, 2018), 403-421. https://doi:10.1111/socf.12422.

248 Rome, Adam. "Earth Day 1970 Was More Than a Protest. It Built a Movement." *Washington Post*, April 22, 2020. https://www.washingtonpost.com/outlook/2020/04/22/earth-day-1970-was-more-than-protest-it-built-movement/.

249 De Moor, Joost, Michiel De Vydt, et al. "New Kids on the Block: Taking Stock of the Recent Cycle of Climate Activism." Taylor & Francis, October 28, 2020. https://www.tandfonline.com/doi/full/10.1080/14742837.2020.1836617?scroll=top&needAccess=true.

250 Fisher, Dana R. "COP-15 in Copenhagen: How the Merging of Movements Left Civil Society Out in the Cold." *Global Environmental Politics* 10, no. 2 (May 1, 2010): 11-17. https://doi.org/10.1162/glep.2010.10.2.11.

251 Ibid.

252 Fisher, Dana R., and Sohana Nasrin. "Climate Activism and Its Effects." *Wiley Interdisciplinary Reviews: Climate Change* 12, no. 1 (October 18, 2020). https://doi.org/10.1002/wcc.683.

253 Moor, Joost de, Michiel De Vydt, et al. "New Kids on the Block: Taking Stock of the Recent Cycle of Climate Activism." *Social Movement Studies* 20, no. 5 (October 28, 2020): 619-625. https://doi.org/10.1080/14742837.2020.1836617

254 Axelrod, Joshua. "Corporate Honesty and Climate Change: Time to Own up and Act." NRDC, February 27, 2019. https://www.nrdc.org/experts/josh-axelrod/corporate-honesty-and-climate-change-time-own-and-act.

255 Bender, Albert. "The Landback Movement Is Decolonizing Indigenous Land across the Americas." *People's World*, January 21, 2022. https://www.peoplesworld.org/article/the-landback-movement-is-decolonizing-indigenous-land-across-the-americas/.

256 Tannous, Nadya, May 17, 2022.

257 Adams, Dawn Neptune, Maulian Dana, and Adam Mazo. "New England Once Hunted and Killed Humans for Money. We're Descendants of the Survivors." *The Guardian*, November 15, 2021. https://www.theguardian.com/commentisfree/2021/nov/15/new-england-once-hunted-and-humans-for-money-were-descendents-of-the-survivors.

258 Pruitt, Sarah. "Treaties Brokered-and Broken-with Native American Tribes." History.com. A&E Television Networks, November 10, 2020. https://www.history.com/news/native-american-broken-treaties.

259 Schuster, Richard, Ryan R. Germain, and Joseph R. Bennett. "Vertebrate Biospanersity on Indigenous-Managed Lands in Australia, Brazil, and Canada Equals That in Protected Areas." Edited by Nicholas J. Reo and Peter Arcese. *Environmental Science & Policy* 101 (November 2019): 1-6. https://doi.org/10.1016/j.envsci.2019.07.002.

260 Hoffman, Kira M., Emma L. Davis, et al. "Conservation of Earth's Biospanersity Is Embedded in Indigenous Fire Stewardship." *Proceedings of the National Academy of Sciences* 118, no. 32 (August 10, 2021). https://doi.org/10.1073/pnas.2105073118.

261 Indigenous Resistance Against Carbon. 2021. Oil Change International.

262 Bullard, Robert D. "Solid Waste Sites and the Black Houston Community*." *Sociological Inquiry* 53, no. 2-3 (April 1983): 273-88. https://doi.org/10.1111/j.1475-682X.1983.tb00037.x.

263 Berndt, Brooks. "30th Anniversary: The First National People of Color Environmental Leadership Summit." United Church of Christ, March 25, 2021. https://www.ucc.org/30th-anniversary-the-first-national-people-of-color-environmental-leadership-summit/.

264 US Environmental Protection Agency. "Environmental Justice Timeline." April 15, 2015. https://www.epa.gov/environmentaljustice/environmental-justice-timeline.

265 Chow, Denise. " 'Transformational': Environmental Justice Advocates See Hope in Debate Discussion." *NBC News*, October 23, 2020. https://www.nbcnews.com/news/us-news/transformational-environmental-justice-advocates-see-hope-debate-discussion-n1244607.

266 White House Council on Environmental Quality. "Response by the White House Council on Environmental Quality to the White House Environmental Justice Advisory Council's Final Recommendations: Justice40, Climate And Economic Justice Screening Tool, And Executive Order 12898 Revisions That Were Submitted on May 21, 2021," May 20, 2022. https://www.epa.gov/system/files/documents/2022-05/CEQ_Response_to_the_WHEJAC_May_2021_Recommendations.pdf.

267 Thomas, Leah. Instagram post, May 28, 2020. https://www.instagram.com/p/CAvaxdRJRxu/?utm_source=ig_embed&ig_rid=3dc77f41-0a06-477f-9eb5-00c7a43f60c4.

268 Crenshaw, Kimberle. "Demarginalizing the Intersection of Race and Sex: A Black Feminist Critique of Antidiscrimination Doctrine, Feminist Theory and Antiracist Policies." University of Chicago Legal Forum 1989, no. 1 (1989): 139-167.

269 Intersectional Environmentalist. "The IE Pledge." Accessed June 30, 2022. https://www.intersectionalenvironmentalist.com/take-the-pledge.

270 Katz, Sabs, April 27, 2022.

271 ClientEarth Communications. "Fossil Fuels and Climate Change: The Facts." *ClientEarth*, February 18, 2022. https://www.clientearth.org/latest/latest-updates/stories/fossil-fuels-and-climate-change-the-facts/.

272 Fossil Fuel Non-Proliferation Treaty. "Who Has Endorsed?" Accessed June 30, 2022. https://fossilfueltreaty.org/endorsements.

273 Campbell, Troy H., and Aaron C. Kay. "Solution Aversion: On the Relation Between Ideology and Motivated Disbelief." *Journal of Personality and Social Psychology* 107, no. 5 (2014): 809-24. https://doi.org/10.1037/a0037963.

274 SIPA Center on Global Energy Policy. "What You Need to Know About a Federal Carbon Tax in the United States." Columbia University. Accessed June 30, 2022. https://www.energypolicy.columbia.edu/what-you-need-know-about-federal-carbon-tax-united-states.

275 Anderson, Soren, Ioana Elena Marinescu, and Boris Shor. "Can Pigou at the Polls Stop US Melting the Poles?" June 4, 2022.

276 Baldwin, Matthew, and Joris Lammers. "Past-Focused Environmental Comparisons Promote Proenvironmental Outcomes for Conservatives." *Proceedings of the National Academy of Sciences* 113, no. 52 (December 27, 2016): 14953-14957. https://doi.org/10.1073/pnas.1610834113.

277 The American Climate Contract. "Supporters in the House of Representatives." Accessed July 1, 2022. https://www.climatesolution.eco/house.

278 "Conservative Climate Caucus." Congressman Curtis. Accessed July 1, 2022. https://curtis.house.gov/conservative-climate-caucus/.

279 Saha, Devashree, and Joel Jaeger. "Ranking 41 US States Decoupling Emissions and GDP Growth." World Resources Institute, July 28, 2020. https://www.wri.org/insights/ranking-41-us-states-decoupling-emissions-and-gdp-growth.

280 Center for Climate and Energy Solutions. "State Climate Policy Maps." December 7, 2021. https://www.c2es.org/content/state-climate-policy/.

281   Turner, Chelsea, and American Clean Power. "Clean Power Annual Market Report 2021." May 17, 2022. https://cleanpower.org/market-report-2021/.

282   Backer, Benjamin, May 24, 2022.

CHAPTER 10: CHANGE IN THE MARKETS

283   Smith, N. Craig. "Consumer Boycotts." *Management Decision* 27, no. 6 (June 1, 1989). https://doi.org/10.1108/EUM0000000000050.

284   Little, Becky. "Key Steps That Led to End of Apartheid." History, November 23, 2020. https://www.history.com/news/end-apartheid-steps.

285   Slaper, Timothy F., and Tanya J. Hall. "The Triple Bottom Line: What Is It and How Does It Work?" *Indiana Business Review* 86, no. 1 (2011): 4–8. https://doi.org/http://www.ibrc.indiana.edu/ibr/2011/spring/article2.html.

286   Business Roundtable. "Our Commitment." Business Roundtable, August 19, 2019. https://opportunity.businessroundtable.org/ourcommitment/.

287   Fink, Larry. "Larry Fink's Annual 2022 Letter to CEOs." BlackRock, 2022. https://www.blackrock.com/corporate/investor-relations/larry-fink-ceo-letter.

288   TED. "The Case for Optimism on Climate Change." Video. *YouTube*, March 14, 2016. https://www.youtube.com/watch?v=gVfgkFaswn4.

289   IEA. "Renewables 2020." 2020. https://www.iea.org/reports/renewables-2020.

290   IRENA. "Majority of New Renewables Undercut Cheapest Fossil Fuel on Cost." June 22, 2021. https://www.irena.org/news/pressreleases/2021/Jun/Majority-of-New-Renewables-Undercut-Cheapest-Fossil-Fuel-on-Cost.

291   Bhutada, Govind. "Breaking Down the Cost of an EV Battery Cell." *Visual Capitalist*, February 22, 2022. https://www.visualcapitalist.com/breaking-down-the-cost-of-an-ev-battery-cell/.

292   Boudette, Neal E., and Coral Davenport. "G.M. Will Sell Only Zero-Emission Vehicles by 2035." *New York Times*, January 28, 2021. https://www.nytimes.com/2021/01/28/business/gm-zero-emission-vehicles.html#:~:text=General%20Motors%20said%20Thursday%20that,trucks%20and%20sport%20utility%20vehicles.

293   Gearino, Dan. "Inside Clean Energy: US Battery Storage Soared in 2021, Including These Three Monster Projects." *Inside Climate News*, March 31, 2022. https://insideclimatenews.org/news/31032022/inside-clean-energy-battery-storage/.

294   Lewis, Michelle. "Solar and Battery Storage Make up 60% of Planned New US Electric Generation Capacity." *Electrek*, March 7, 2022. https://electrek.

295   co/2022/03/07/solar-and-battery-storage-make-up-60-of-planned-new-us-electric-generation-capacity/.

295   International Renewable Energy Agency. "Renewable Energy and Jobs Annual Review 2020." October 2021. https://www.irena.org/publications/2021/Oct/Renewable-Energy-and-Jobs-Annual-Review-2021.

296   American Clean Power. "2022 Q1 Clean Power Quarterly Market Report," 2022. https://cleanpower.org/wp-content/uploads/2022/05/2022_CPQReport_Q1_Public-Version.pdf.

297   ClientEarth. "Fossil Fuels and Climate Change: The Facts," February 18, 2022. https://www.clientearth.org/latest/latest-updates/stories/fossil-fuels-and-climate-change-the-facts/#:~:text=In%202018%2C%2089%25%20of%20global,source%20of%20global%20temperature%20rise.

298   Jacobson, Mark Z., Anna-Katharina von Krauland, et al. "Low-Cost Solutions to Global Warming, Air Pollution, and Energy Insecurity for 145 Countries." *Energy & Environmental Science* 15, no. 8 (June 28, 2022): 3343-59. https://doi.org/10.1039/D2EE00722C.

299   Whitmarsh, Lorraine. "Behavioural Responses to Climate Change: Asymmetry of Intentions and Impacts." *Journal of Environmental Psychology* 29, no. 1 (March 2009): 13-23. https://doi.org/10.1016/j.jenvp.2008.05.003.

300   Accenture. "COVID-19: New Retail Consumer Behavior Habits." *Accenture*, August 13, 2020. https://www.accenture.com/us-en/insights/retail/coronavirus-consumer-habits.

301   Bloomberg and American Petroleum Institute. "Earnings in Perspective." 2022. https://www.api.org/oil-and-natural-gas/energy-primers/earnings-in-perspective.

302   Climate Safe Pensions and Stand.earth. "The Quiet Culprit: Pension Funds Bankrolling the Climate Crisis," n.d. https://climatesafepensions.org/wp-content/uploads/2021/12/CSPN-The-Quiet-Culprit.pdf.

303   Gray, Amy, May 12, 2022.

304   Global Fossil Fuel Commitments Database. "Global Fossil Fuel Commitments Database." 2022. https://divestmentdatabase.org/.

305   Boffey, Daniel. "One Of World's Biggest Pension Funds To Stop Investing In Fossil Fuels." 2021. *The Guardian*. https://www.theguardian.com/environment/2021/oct/26/abp-pension-fund-to-stop-investing-in-fossil-fuels-amid-climate-fears.

306   Barnard, Anne. "New York's $226 Billion Pension Fund Is Dropping Fossil Fuel Stocks." 2020. *New York Times*. https://www.nytimes.com/2020/12/09/nyregion/new-york-pension-fossil-fuels.html.

307　Callahan, Mary. "California May Require Pension Divestment From Fossil Fuels." 2022. *Governing.* https://www.governing.com/now/california-may-require-pension-divestment-from-fossil-fuels.

308　Fairfax, Lisa M. "Social Activism Through Shareholder Activism." Washington and Lee University School of Law Scholarly Commons, February 15, 2019. https://scholarlycommons.law.wlu.edu/wlulr/vol76/iss3/3/.

309　Krueger, Philipp, Zacharias Sautner, and Laura T. Starks. "The Importance of Climate Risks for Institutional Investors." *Swiss Finance Institute Research Paper,* no. 18-58 (November 11, 2019).

310　Follow This. "Resolution at 2016 AGM of Royal Dutch Shell Plc ('Shell')." 2016. https://www.follow-this.org/wp-content/uploads/2020/04/Follow-This-shareholder-resolution-Shell-2016-full-text.pdf.

311　Blackburne, Alex. "Activist Investors Turning up Heat on Oil Majors in Proxy Voting Season." *S&P Global Market Intelligence,* April 25, 2022. https://www.spglobal.com/marketintelligence/en/news-insights/latest-news-headlines/activist-investors-turning-up-heat-on-oil-majors-in-proxy-voting-season-69438831.

312　Khaliq, Sehr. "Never Underestimate the Power of the Proxy." Interfaith Center on Corporate Responsibility, 2022. https://www.iccr.org/never-underestimate-power-proxy.

313　Blackburne, "Activist Investors Turning up Heat on Oil Majors in Proxy Voting Season."

CHAPTER 13: ECONOMIES OF PEACE—PAKISTAN

314　Miles-Novelo, Andreas, and Craig A. Anderson. *Climate Change and Human Behavior: Impacts of a Rapidly Changing Climate on Human Aggression and Violence.* Elements in Applied Social Psychology. Cambridge: Cambridge University Press, 2022. doi:10.1017/9781108953078.

Mango Publishing, established in 2014, publishes an eclectic list of books by diverse authors—both new and established voices—on topics ranging from business, personal growth, women's empowerment, LGBTQ+ studies, health, and spirituality to history, popular culture, time management, decluttering, lifestyle, mental wellness, aging, and sustainable living. We were named 2019 *and* 2020's #1 fastest growing independent publisher by *Publishers Weekly*. Our success is driven by our main goal, which is to publish high-quality books that will entertain readers as well as make a positive difference in their lives.

Our readers are our most important resource; we value your input, suggestions, and ideas. We'd love to hear from you—after all, we are publishing books for you!

Please stay in touch with us and follow us at:

Facebook: Mango Publishing
Twitter: @MangoPublishing
Instagram: @MangoPublishing
LinkedIn: Mango Publishing
Pinterest: Mango Publishing
Newsletter: mangopublishinggroup.com/newsletter

Join us on Mango's journey to reinvent publishing, one book at a time.